D1522730

Carbonate Reservoir Characterization

F. Jerry Lucia

Springer

Berlin
Heidelberg
New York
Barcelona
Hong Kong
London
Milan
Paris
Singapore
Tokyo

F. Jerry Lucia

Carbonate Reservoir Characterization

With 171 Figures and 5 Tables

 Springer

F. Jerry Lucia
Senior Research Fellow
University of Texas at Austin
University Station
Box X
Austin, Texas 78713-8924

ISBN 3-540-63782-6 Springer-Verlag Berlin Heidelberg New York

Library of Congress Cataloging-in-Publication Data

Lucia, F. Jerry.
Carbonate reservoir characterization / F. Jerry Lucia.
p.   cm.
Includes bibliographical references and Index.
ISBN 3-540-63782-6 (hardcover)
1. Carbonate reservoirs. 2. Oil reservoir engineering.
I. Title.
TH871.L83   1999
622'.3382–dc21   98-29500

© Springer-Verlag Berlin Heidelberg 1999
Printed in Germany

The use of general descriptive names, registered names, trademarks, etc. in this publication does not imply, even in the absence of a specific statement, that such names are exempt from the relevant protective laws and regulations and therefore free for general use.

Cover: Erich Kirchner, Heidelberg
Typesetting: Data conversion by MEDIO, Berlin

SPIN: 10634893   32/3020 - 5 4 3 2 1 0 – Printed on acid-free paper.

Preface

This book is intended for geologists and engineers interested in carbonate reservoir characterization. Major advances in the field of carbonate reservoir geology have been made in the past 10 years, brought about by the need for more realistic detailed reservoir models for input into highly advanced computer fluid-flow simulators. The purpose of this book is to summarize advanced methods used in constructing realistic carbonate reservoir models. Material included in this book has been used by the author as a textbook for courses on carbonate reservoir characterization. The author hopes that others will find it useful for learning about carbonate reservoirs and as a source for additional information on the subject.

Reservoir model construction requires the integration of geological and engineering data, and methods of accomplishing this difficult task are at the center of the book. Reservoir characterization can be defined in many different ways, but here the focus is on methods of distributing petrophysical properties in three-dimensional space using geologic models. Therefore, the first chapter is a review of petrophysical properties, Chapter 2 discusses the link between petrophysical properties and rock fabrics, and Chapter 3 presents empirical methods of obtaining rock-fabric information from wireline logs. Chapters 4 through 7 focus on using depositional and diagenetic processes to predict the spatial distribution of rock-fabric facies. Chapter 8 discusses methods of constructing realistic reservoir models using the rock-fabric approach.

I wrote this book because of new developments in the procedure for integrating geologic and engineering data, a task that is key to any reservoir characterization project. The relationship between geology and petrophysical properties has been the subject of intense study since 1955. Previously the study of carbonate rocks was the exclusive realm of the paleontologists; stratigraphy was "biostratigraphy" and oil fields were "reefs." During the 1950s, investigations of modern carbonate sediments demonstrated the importance of carbonate petrography and sedimentation, studies that started with Leslie Illing's paper on the Bahama Platform and that developed into a sound basis for application to ancient carbonate units, principally through the efforts of Robert Ginsburg. During this period, Gus Archie demonstrated that carbonate reservoirs can be characterized by petrophysical properties and can be described as porous layers instead of as "reefs." During the 1960s Ray Murray, Perry Roehl, and the author published the first efforts to link petrophysical properties to carbonate petrography and sedimentation through detailed geological and petrophysical field studies using core data. This link is vital for predicting the spatial distribution of petrophysical properties.

As the number of geological studies increased in the 1970s, depositional facies became the principal link to porosity and permeability, and diagenetic studies were important for determining which depositional facies were reservoir facies. The concept that depositional facies had predictable patterns was fully developed by Jim Wilson in 1975. Constructing facies models has been the principal method for characterizing carbonate reservoirs and predicting the 3-D distribution of petrophysical properties in the 1970s and 1980s. This approach was summarized by a compilation of field studies edited by Perry Roehl and Philip Choquette in 1985.

The link between depositional facies and petrophysical properties was never well developed. Depositional facies are commonly described in terms of allochems, biota, and texture, such as fusulinid/ooid packstone/grainstone. These facies descriptions focus on the interpretations of depositional environment and have little direct relation to petrophysical properties. As carbonate reservoirs in the United States and Canada matured, more detailed petrophysical models were required for predicting recovery from redevelopment programs. Attempts to simulate reservoir performance using computer simulation programs developed by reservoir engineers during the 1970s led to the conclusion that simple petrophysical models were inadequate to describe reservoirs. What was needed was a more heterogeneous pattern. To add heterogeneity to facies models, the engineering community used newly developed geostatistical methods, such as variography. The resulting models were not realistic because depositional facies are not sufficiently detailed to provide the framework needed to describe petrophysical heterogeneity.

New methods were needed for constructing a geologic framework and linking facies to petrophysical properties. The advent of carbonate sequence stratigraphy, principally through the efforts of Charles Kerans, has provided the stratigraphic framework required for describing reservoir heterogeneity in geologic terms. Sequence stratigraphy defines chronostratigraphic surfaces that are continuous throughout the reservoir, whereas depositional textures typically are not continuous. Vertical successions and lateral progressions of depositional textures are systematically related to accommodation space, current energy, topography, and biologic activity. Depositional textures can be distributed between chronostratigraphic surfaces using these systematic relationships.

Rock-fabric studies provided the improved link between geology and petrophysics by focusing on current fabrics. This link has been known for some time but only recently has been fully developed to fill the need for a more detailed petrophysical model. Included in these developments are empirical methods for obtaining rock-fabric information from wireline logs and integrating rock-fabric information into wireline log calculations. Rock fabrics are the products of both depositional and diagenetic processes. The distribution of diagenetic products is a function of the precursor and geochemical-hydrological conditions, which are poorly understood. Thus, predicting the distribution of petrophysical properties is difficult when diagenetic products do not conform to depositional patterns.

This book is organized around these new methods for constructing a geologic framework and linking facies to petrophysical properties. Petrophysical properties are related to rock fabrics; rock fabrics are related to depositional textures and diagenetic product; depositional textures and diagenetic products are related to diagenetically modified depositional rock-fabric facies; the spatial distribution of depositional facies are predicted using sequence stratigraphic concepts; and reservoir models are constructed using patterns of rock-fabric facies constrained by chronostratigraphic surfaces. This approach, sometimes referred to as the "Archie Method," is offered as containing the fundamental elements for constructing a realistic 3-D reservoir model.

Austin, Texas   F. Jerry Lucia
October 1998

References

Archie GE (1952) Classification of carbonate reservoir rocks and petrophysical considerations. AAPG Bull 36, 2: 278–298
Illing LV (1954) Bahaman calcareous sands. AAPG Bull 38, 1: 1–95
Kerans C, Fitchen WM (1995) Sequence hierarchy and facies architecture of a carbonate-ramp system: San Andres Formation of Algerita Escarpment and western Guadalupe Mountains, West Texas and New Mexico. The University of Texas at Austin, Bureau of Economic Geology Report of Investigation No 235, 86 pp
Lucia FJ, Murray RC (1966) Origin and distribution of porosity in crinoidal rocks. Proc 7th World Petroleum Congress, Mexico City, Mexico, 1966, pp 409–423
Murray RC (1960) Origin of porosity in carbonate rocks. J Sediment Petrol 30: 59–84
Roehl PO (1967) Stony Mountain (Ordovician) and Interlake (Silurian) facies analogs of Recent low-energy marine and subaerial carbonate, Bahamas. AAPG Bull 51, 10: 1979–2032
Roehl RO, Choquette PW (1985) Carbonate petroleum reservoirs. Springer, Berlin Heidelberg New York, 622pp

Contents

Chapter 3
Rock-Fabric/Petrophysical Properties from Core Description and Wireline Logs:
The One-Dimensional Approach

Chapter 4
Origin and Distribution of Depositional Textures and Petrophysical Properties:
The Three-Dimensional Approach

Chapter 5
Diagenetic Overprinting and Rock-Fabric Distribution:
The Cementation, Compaction, and Selective Dissolution Environment

Chapter 6
Diagenetic Overprinting and Rock-Fabric Distribution:
The Dolomitization/Evaporite-Mineralization Environment

Chapter 7
Diagenetic Overprinting and Rock-Fabric Distribution:
The Massive Dissolution, Collapse and Fracturing Environment

Chapter 8
Reservoir Models for Input into Flow Simulators

CHAPTER 1

Petrophysical Rock Properties

1.1
Introduction

The principal goal of reservoir characterization is to construct three- and four-dimensional images of petrophysical properties. The purpose of this chapter is to review basic definitions and laboratory measurements of the petrophysical properties porosity, permeability, relative permeability, capillarity, and saturation. Pore-size distribution is presented as the common link between these properties.

1.2
Porosity

Porosity is an important rock property because it is a measure of the potential storage volume for hydrocarbons. Porosity in carbonate reservoirs ranges from 1 to 35% and, in the United States, averages 10% in dolomite reservoirs and 12% in limestone reservoirs (Schmoker et al. 1985).

Porosity is defined as pore volume divided by bulk volume.

$$\text{Porosity} = \frac{\text{Pore volume}}{\text{Bulk volume}} = \frac{\text{Bulk volume - Mineral volume}}{\text{Bulk volume}} \qquad (1)$$

Fractional porosity is used in engineering calculations. Geologists, however, most commonly refer to porosity as a percentage (porosity x 100). The term "effective porosity" or "connected" pore space is commonly used to denote porosity that is most available for fluid flow. However, at some scale all pore space is connected. The basic question is how the pore space is connected.

Porosity is a scalar quantity because it is a function of the bulk volume used to define the sample size. Therefore, whereas the porosity of Carlsbad Caverns is 100% if the caverns are taken as the bulk volume, the porosity of the formation including Carlsbad Caverns is much less and depends upon how much of the surrounding formation is included as bulk volume.

Porosity is determined by visual methods and laboratory measurements. Visual methods of measuring total porosity are estimates at best because the amount of porosity visible depends on the method of observation: the higher the magnification the more pore space is visible, for example. Porosity is commonly estimated by visual inspection of core slabs using a low power microscope. The Archie classification (Archie 1952) provides a method of combining textural criteria and visible porosity to determine total porosity. Visible porosity can be measured by point counting the porosity seen in thin sections or by using image analysis software to calculate pore space seem in images of thin sections. Visual estimates can be very inaccurate unless calibrated against point-counted values. Commonly, visual estimates of porosity are twice as high as point-counted values.

Measurement of porosity of rock samples in the laboratory requires knowing the bulk volume of the rock and either its pore volume or the volume of the matrix material (grain volume). Bulk volume is usually measured by volumetric displacement of a strongly nonwetting fluid, such as mercury, or by direct measurement. The accuracy of laboratory measurements depends upon the method used and the care taken. The least accurate method of determining pore volume is the summation of fluids. The most accurate method in use is a method based on gas expansion and Boyle's gas law. The injection of mercury under very high pressure is also used to measure porosity as well as pore-size distribution. Laboratory porosity values are normally higher than visible porosity because very small pore space cannot be seen with visual observational techniques. When all the pore space is large and visible, however, visible porosity is comparable with measured values.

Inaccuracies in laboratory measurements arise from poor sample preparation and sampling procedures. Problems in sample preparation include (1) incomplete removal of all fluids and (2) alteration of rock fabrics that contain minerals with bound water such as gypsum and clay minerals. An example of incomplete removal of all fluids is taken from the Seminole San Andres field, where samples were cleaned a second time, increasing the porosity values 0–4 porosity-percent (Fig. 1). Subjecting samples that contain gypsum to high temperatures in laboratory procedures results in dehydrating gypsum to the hemihy-

Fig. 1. Plot of whole-core porosity values versus porosity values of plug samples taken from the whole-core samples and recleaned. Whole-core porosity is too small by 0-4 porosity percent

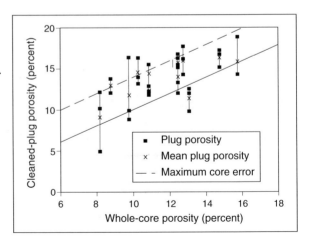

Fig. 2. Effect of confining pressure on porosity in Paleozoic and Jurassic carbonate reservoir. Porosity loss is defined as *confined porosity/ unconfined porosity*

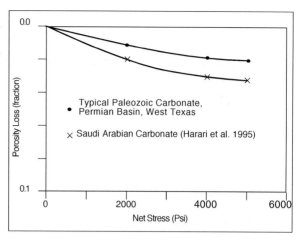

drate form Bassanite (see Eq. 2) and creating pore space and free water, both of which produce large errors in the porosity measurements (Table 1). Pore space is created because Bassanite has a smaller molar volume than gypsum.

$$\text{Gypsum } (CaSO_4 2H_2O) + \text{heat } = \text{ Bassanite } (CaSO_4 \cdot 0.5H_2O) + 1.5H_2O \quad (2)$$
$$\text{(Density 2.35)} \qquad\qquad\qquad \text{(Density 2.70)}$$

Porosity measurements should be made at in situ stress conditions because carbonate rocks are compressible, and porosity decreases with increasing effective stress. The common laboratory method is to increase confining pressure while maintaining a constant pore pressure. The resulting decrease in porosity is normally very small (2%) in Paleozoic and many Mesozoic age reservoirs (Fig. 2), and porosity measurements at ambient conditions are usually adequate (Harari et al. 1995). Porosity values of all high-porosity carbonates, however, should be checked for porosity loss with increasing confining pressure.

Table 1. Increase in porosity of gypsum-bearing dolomite samples due to heating (After Hurd and Fitch 1959)

	Low temp. analysis	High temp. analysis	Increase in porosity
Gypsum (%)	Porosity (%)	Porosity (%)	Porosity (%)
4.3	2.8	3.7	0.9
14.6	2.5	8.4	5.9
14.9	3.4	8.9	5.5
11.0	6.4	11.2	4.7

Fig. 3. Comparison of porosity in **A** cubic packed spheres and **B** rhombohedral-packed spheres. The porosity is a function of packing, and pore size is controlled by the size and packing of sphere

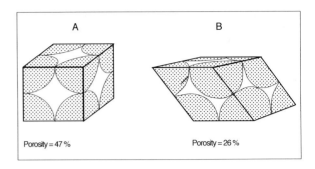

The common pore type in sedimentary rocks is intergrain. The percent intergrain porosity in cubic packed spherical grains can be calculated to be 47.6% (Fig. 3A). Cubic packing is the most open arrangement of grains. Rhombohedral packing is the closest arrangement of spherical grains and has a porosity of 25.9 percent (Fig. 3B). The effect of packing can be seen in unconsolidated sandstone. The porosity of extremely well sorted unconsolidated sandstone is 42% whereas the porosity of poorly sorted sandstone is 27% (Beard and Weyl 1973). In unconsolidated carbonate sediment the shape of the grains and intragrain porosity as well as grain packing have a large effect on porosity. The porosity of modern ooid grainstones is 45% but, in contrast to siliciclastics, the porosity of carbonate sediments increases to 70 percent as sorting decreases (Enos and Sawatski 1981). This increase is related to pore space located within carbonate grains and to the needle shape of the mud-sized aragonite crystals. Diagenetic processes, such as cementation and compaction, reduce porosity to very low values.

Although intergrain porosity is not a function of grain size, the size of intergrain pores is a function of grain size, sorting, and intragrain porosity (Lucia 1995). Pore size decreases as grain size and sorting decreases. In addition, the intergrain pore size is reduced systematically as cement occludes the pore space or as grains are forced closer together by compaction.

1.3
Permeability

Permeability is important because it is a rock property that relates to the rate at which hydrocarbons can be recovered. Values range considerably from less than 0.01 md to well over 1 Darcy. A permeability of 0.1 md is generally considered minimum for oil production. Highly productive reservoirs commonly have permeability in the Darcy range.

Permeability is expressed in Darcy's Law:

$$\text{Darcy's Law:} \quad Q = K\left(\frac{I}{M}\right)\left(\frac{P_1 - P_2}{L}\right)A \tag{3}$$

Where Q is rate of flow, K is permeability, μ is fluid viscosity, $(P_1 - P_2)/L$ is the potential drop across a horizontal sample, and A is the cross-sectional area of the sample.

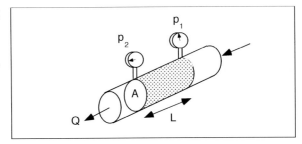

Fig.4. Method of measuring permeability of a core plug in the laboratory. Samples are oriented horizontally to eliminate gravity effects; see text for explanation

Permeability is a rock property, viscosity is a fluid property, and $\Delta p/L$ is a measure of flow potential. Permeability is measured in the laboratory by encasing the sample in an air-tight sleeve and flowing a fluid of known viscosity through a sample of known length and diameter while mounted in a horizontal position (Fig. 4). The pressure drop across the sample and flow rate are measured and permeability calculated. Normally, either air or brine is used as a fluid and, when high rates of flow can be maintained, the results are comparable. At low rates, air permeability will be higher than brine permeability. This is because gases do not adhere to the pore walls as do liquids, and the slippage of gases along the pore walls gives rise to an apparent dependence of permeability on pressure. This is called the Klinkenberg effect, and it is especially important in low permeability rocks. Carbonate samples often contain small fractures, which maybe natural or induced, and stylolites, which are natural. In unconfined conditions, these features tend to be flow channels and may result in unreasonably high permeability values, Therefore, measurements should be made when the sample is under some confining pressure, preferably a confining pressure equivalent to in situ reservoir conditions. The reduction in permeability with increasing stress, however, is normally small in Paleozoic and Mesozoic reservoirs.

A measure of permeability is obtained through a device called the miniairpermeameter (Hurst and Goggin 1995). This device is designed for outcrop measurements but is also used in the laboratory to obtain numerous permeability values economically. It consists of a pressure tank, a pressure gage, and a length of plastic hose with a special nozzle or probe designed to fit snugly against the rock sample. The gas pressure and the flow rate into the rock sample are used to calculate permeability.

Permeability is a vector and scalar quantity. Vertical permeability is commonly about half of the horizontal permeability, and horizontal permeability varies with the direction measured. Core analysis reports commonly give a maximum permeability and a permeability value measured at 90 degrees from maximum. Darcy's Law shows permeability as a function of the cross-sectional area of the sample. Permeability, therefore, depends upon the area of investigation. In carbonate reservoirs, permeability is highly variable on the scale of inches and feet and permeability variability may reflect the size and location of the sample. The permeability of a small sample may not be representative of the formation the

Fig. 5. Typical Horner pressure buildup plot (after Dake 1978). The slope of the line is a function of permeability-feet (kh)

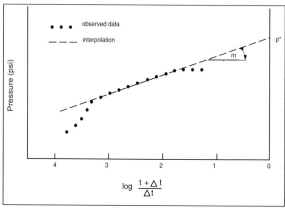

sample came from. Therefore, it is important to know how the formation was sampled before using the laboratory results. *Never use core data unless you have seen the core and observed how the core was sampled.*

A measure of permeability can be obtained from wells because reservoir pressure is reduced when fluid is produced in proportion to the rate of production. This relationship is used to calculate fluid transmissibility using the rate of pressure change and production volumes measured from test intervals in wells. Fluid transmissibility is expressed as permeability-feet (kh), and average permeability values are commonly obtained by dividing kh by the vertical height of the test interval. A large error in average permeability can result if the proper height is not known.

Pressure buildup tests are used to calculate the reservoir pressure, the effective permeability, and well bore damage (skin effect, Fig. 5). The well is flowed at a constant rate for a time (T) and then shut in. During the shut-in period (ΔT) the increase in shut-in pressure is recorded. Flow rates and pressure changes are analyzed using the Horner pressure buildup plot, which is a graph of shut-in pressure versus the log of dimensionless time, $(T+\Delta T)/\Delta T$. Reservoir pressure is determined by extrapolating the straight-line portion to 0 (log of 1). The skin calculation is related to the deviation of the buildup curve during the initial pressure buildup.

The effective, average permeability of the interval tested is calculated using the following equation.

$$\text{Slope (psi/log cycle)} = 162.6(q\mu B_o/kh), \tag{4}$$

Where q is the flow rate in stock-tank-barrels/day, μ = viscosity in centipoises, B_o is reservoir-barrels/stock-tank-barrels, k is permeability in millidarcies, and h is the net reservoir interval in feet.

It is common practice to estimate permeability using simple porosity-permeability transforms developed from core data. However, porosity-permeability cross plots for carbonate reservoirs commonly show large variability (Fig. 6),

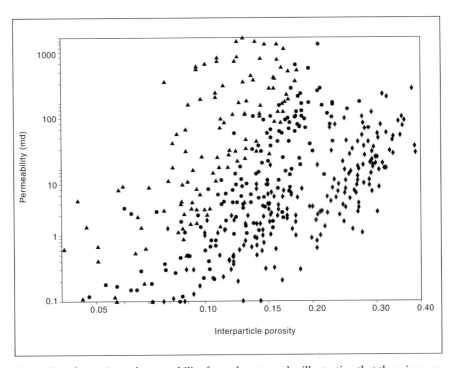

Fig. 6. Plot of porosity and permeability for carbonate rocks, illustrating that there is no relationship between porosity and permeability in carbonate rocks without including pore-size distribution

demonstrating that factors other than porosity are important in modeling permeability. Other factors are the size and distribution of pore space, or pore-size distribution. *There is no relationship between porosity and permeability in carbonate rocks unless pore-size distribution is included.*

Permeability models have historically described pore space in terms of the radius of a series of capillary tubes. The number of capillary tubes has been equated to porosity so that permeability is a function of porosity and pore-radius squared (Eq. 5 from Amyx, et al. 1960, p. 97). Kozeny (1927) substituted surface area of the pore space for pore radius and developed the Kozeny equation relating permeability to porosity, surface-area squared, and the Kozeny constant (Eq. 6).

$$k = \pi r^2/32, \text{ or } k = \phi r^2/8 \tag{5}$$

$$k = \phi/k_z S_p^{\,2} \tag{6}$$

As discussed in Section 1.2, intergrain pore size, or pore radius, is a function of grain size, sorting, and intergrain porosity, a function that can be used to relate permeability to rock fabrics. Capillary pressure curves are used to measure pore radius in the laboratory.

1.4
Capillary Properties and Fluid Distribution

Capillary pressure measurements are used to describe pore-size and fluid distribution. The two principal methods of measuring capillary properties are the injection of mercury (nonwetting phase) into a sample containing air (wetting phase) under pressure and the displacement of water (wetting phase) by oil (nonwetting phase) in a centrifuge. The volume of mercury or oil injected at increments of increasing pressure is measured as a fraction of pore volume (saturation), and a plot of injection pressure against saturation is made (Fig. 7). This graph is referred to as the drainage curve. As the injection pressure is reduced, the wetting fluid (air or water) will flow into the pore space and the nonwetting will be fluid expelled. A plot of pressure and saturation during the reduction of injection pressure is referred to as the imbibition curve (Fig. 7). Not all the nonwetting fluid will be expelled for capillary pressure reasons, and the trapped fluid is referred to as residual saturation. This process is called imbibition and is a significant recovery process in water flooding.

Hydrocarbon saturation, expressed as a fraction of pore volume, and hydrocarbon distribution, are dependent on (1) the interfacial tension between hydrocarbons and water, (2) the adhesive forces between the fluids and the minerals that make up the pore walls, (3) the pressure differential between the hydrocar-

Fig. 7. Typical capillary pressure curves showing drainage and imbibition curves. Data for the drainage curve is obtained by increasing pressure, whereas data for the imbibition curve is obtained by reducing pressure

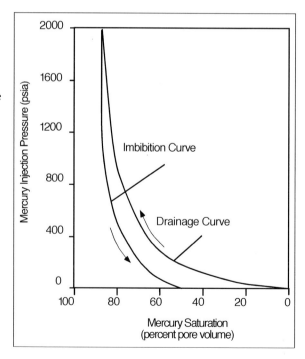

bon and water phases during the process of accumulation (capillary pressure), and (4) pore-size distribution.

Interfacial tension results from the attraction of molecules in a liquid to each other. It can be defined in terms of the pressure across the fluid boundary and the radius of curvature of that boundary, as shown in Eq. (7).

$$2\sigma/rl = (P_1 - P_2), \tag{7}$$

where $P_1 - P_2$ is the pressure differential across the meniscus (capillary pressure), σ is surface tension, and rl is the radius of curvature of the liquid.

Equation (7) can be derived by considering the hemispherical bottom part of a drip of water coming out of a small faucet just before it drops. The molecules within the drip attract each other equally but the molecules on the surface are attracted toward the center of the drip and to the other molecules on the surface, creating a net inward force (Fig. 8).

If $F\downarrow$ = total downward force pulling on drip

then $F\downarrow = \pi r_1^2 (P_1 - P_2),$ (8)

where πr_1^2 = Cross-sectional area of drip and
            $(P_1 - P_2)$ = Difference between the pressure inside the drip (water pressure) and pressure outside the drip (atmospheric pressure).

If $F\uparrow$ = upward cohesive forces holding drip together

Then $F\uparrow = 2\pi r_1 \sigma,$ (9)

where r_1 = radius of curvature of liquid,
            $2\pi r_1$ = circumference of drip, and
            σ = surface tension.

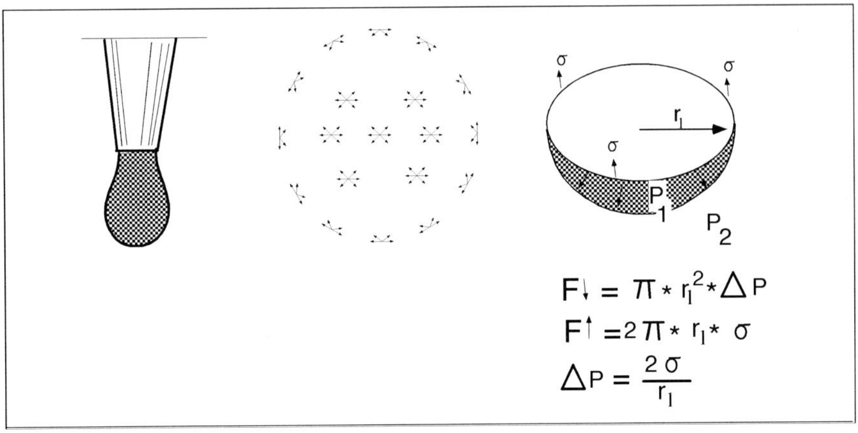

Fig. 8. Cohesive forces and the definition of surface tension

Fig. 9. Adhesive forces and the definition of wettability. If the adhesive forces are less than cohesive forces, ($\theta >$ 90°), the liquid is said to be nonwetting. If adhesive forces are greater than cohesive forces, ($\theta <$ 90°), the liquid is said to be the wetting phase

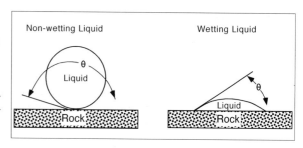

At equilibrium, $F\uparrow = F\downarrow$

$$2\pi \, r_1 \, \sigma = \pi r_1^2 \, (P_1 - P_2)$$

(7)

or

$$2\sigma / r_1 = (P_1 - P_2).$$

Whereas cohesive forces hold the liquid together, adhesive forces between solid and liquid tend to spread the liquid out. When a fluid encounters a solid surface it tends either to spread over that surface or to form a ball on the surface, and the angle between the solid and liquid meniscus is a measure of the adhesive force. The contact angle between the fluid and the surface will be less than 90° for fluids that spread over the surface and greater than 90° for fluids that tend to form a ball. If the contact angle is less than 90°, the fluid is said to be wetting and if greater than 90°, nonwetting (Fig. 9).

The adhesive forces between a solid and water are greater than those between a solid and air in a water/air/solid capillary system and cause water to rise in a capillary tube (Fig. 10). The adhesive force is equal to cos θ, and is equal to the radius of the capillary tube divided by the radius-of-curvature of the liquid meniscus (Eq. 10):

$$\cos \theta = r_c / r_1 , \text{ or } r_1 = r_c / \cos \theta ,$$

(10)

Where r_1 is the radius of the liquid meniscus, r_c is the radius of the capillary tube, and cos θ is the adhesive force (solid/liquid).

A pressure difference (capillary pressure) exists across the air/water interface, and this pressure can be defined in terms of the surface tension (σ) and the radius of curvature of the meniscus (rl):

$$2\sigma / r_1 = (P_1 - P_2) \text{ (see Eq. 1)}.$$

Substituting Eq. (10) for r_1 results in:

$$2\sigma \cos \theta / r_c = (P_1 - P_2) = \text{Capillary Pressure}$$

or:

(11)

$$r_c = 2\sigma \cos \theta / P_c \times 0.145,$$

Fig. 10. Capillary pressure relations in a capillary tube

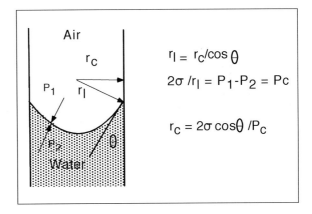

Where r_c is in microns,
         σ is in dynes/cm, and
         P_c is psia (not dynes/cm^2).

Equation (11) is the basic equation used to measure pore-size distribution from a mercury capillary pressure curve. The pores measured are the smaller pores that provide entry into the larger pores, referred to as the pore throats or portals. The pore throat data is presented as a frequency or cumulative frequency plot as shown in Fig. 11.

In order to relate pore-size distribution, determined from mercury capillary pressure data, to permeability and porosity as discussed above, a normalizing pore size must be chosen (Swanson 1981). The pore size at a mercury saturation of 35% has been determined to be most useful (Pittman 1992). Generic equations relating porosity, permeability, and pore throat size at various mercury saturations for siliciclastics have been published by Pittman (1992). A plot using 35% mercury saturation is shown in Fig. 12 and demonstrates that pore throat size has a larger effect on permeability than does porosity. Because these equations are derived for siliciclastic rocks, the porosity is restricted to intergrain porosity and not necessarily to total porosity.

For oil to accumulate in a hydrocarbon trap and form a reservoir, the surface tension between water and oil must be exceeded. This means that the pressure in the oil phase must be higher than the pressure in the water phase. If the pressure in the oil is only slightly greater than that in the water phase, the radius of curvature will be large and the oil will be able to enter only large pores. As the pressure in the oil phase increases the radius of curvature decreases and oil can enter smaller pores (Fig. 13).

In nature, the pressure differential (capillary pressure) is produced by the difference in density between water and oil; the buoyancy effect. At the zero capillary pressure level (zcp), the reservoir pressure is equal to pressure in the water phase (depth × density of water). Above the zcp level the pressure in the water

Fig. 11. Pore throat size distribution plots from mercury capillary pressure curves. **A** Frequency diagram. **B** Cumulative frequency diagram

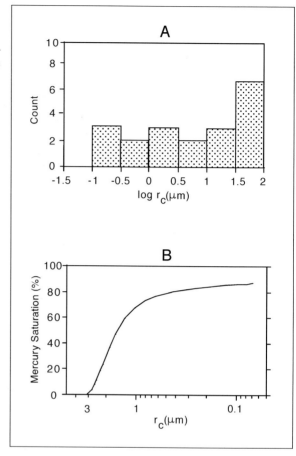

phase will be reduced by the height above the zcp times water density, and the pressure in the oil phase will be reduced by the height above the zcp times oil density.

Pressure in water phase $(P_w) = Pzcp - H\rho_w.$ (12)

Pressure in oil phase $(P_o) = Pzcp - H\rho o.$ (13)

At any height in an oil column, the pressure difference between the oil phase and the water phase (capillary pressure) is the difference between the specific gravity of the two fluids multiplied by the height of the oil column.

$P_o - P_w = P_c = 0.434\ H\ (\rho_w - \rho_o),$ (14)

where H is the height above zero capillary pressure level, ρ_o is the density of the oil phase, ρ_w is the density of the water phase, and 0.434 is the unit conversion constant.

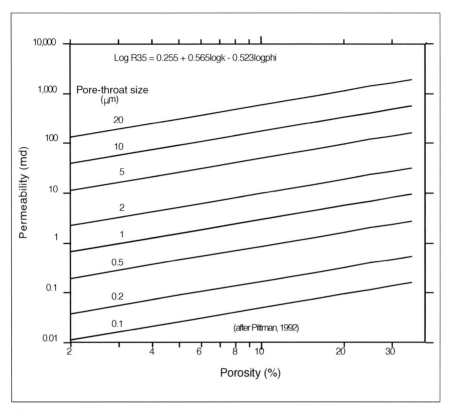

Fig. 12. Plot of porosity, permeability, and pore-throat size for siliciclastic rocks using 35% mercury saturation to calculate pore-throat size from mercury capillary pressure curves (Pittman, 1992)

Mercury capillary pressure measurements can be converted to height above the zero capillary pressure level (reservoir height) by converting the surface tension and contact angle for the fluids used in the laboratory to the values for the specific subsurface fluids in a given reservoir. The equation for this conversion can be derived from capillary theory:

$$h = \frac{(\sigma\cos\theta)o/w/s \times (Pc)hg/a/s}{0.434(\rho_w - \rho_o) \times (\sigma\cos\theta)hg/a/s}, \qquad (15)$$

where o/w/s is the oil/water/solid system, and hg/a/s is the mercury/air/solid system.

Typical values for converting mercury/air capillary pressure curves to reservoir conditions of oil/water are given below in Table 2.

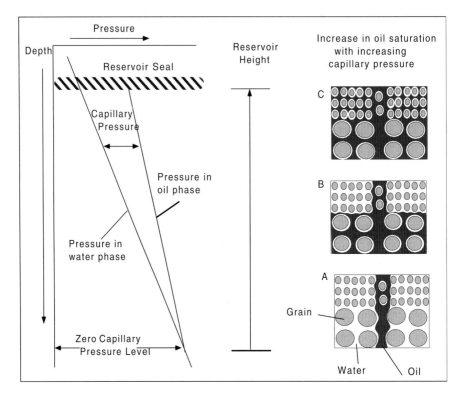

Fig. 13. Diagram showing smaller pores being filled with a non-wetting fluid (oil) displacing a wetting fluid (water) as capillary pressure increases linearly with reservoir height. Pore size is determined by grain size and sorting. (**A**) Only the largest pores contain oil at the base of the reservoir. (**B**) Smaller pores are filled with oil as capillary pressure and reservoir height increase. (**C**) Smallest pores are filled with oil toward the top of the reservoir

Table 2. Typical values for converting mercury/air capillary pressure curves to reservoir conditions of oil/water

LABORATORY mercury/air/solid	RESERVOIR oil/water/solid	RESERVOIR density (g/cc)	
σ 480 dynes/cm	σ 28 dynes/cm	Water(ρ_w)	1.1
θ 140°	θ 33 – 55 Degrees	Oil(ρ_o)	0.8

The difference in density between water and hydrocarbons, the buoyancy force, produces pressures in hydrocarbon columns that exceed pressures in the water column. The pressure gradient in reservoirs can be used to determine the distance above the zcp level. Multiple reservoir pressures can be obtained from the repeat formation tester, a wireline formation tester capable of multiple settings downhole (Smolen and Litsey 1977). It can retrieve several fluid samples

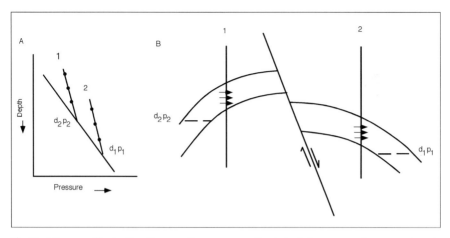

Fig. 14. Diagram illustrating the use of pressure gradients to define reservoir compartments and water levels. (**A**) Depth plot of pressure from wells 1 and 2 shown in (**B**). Intersections of depth plots with the regional fluid-pressure gradient are at depths d_1 and d_2 suggesting separate reservoirs with different water-oil contacts. (**B**) Cross section showing well locations, sampling depths, and sealing fault dividing the hydrocarbon accumulation into two reservoirs

per trip, but its primary advantage is its multiple-level pressure-measuring capability. This tool allows a number of pressures to be taken from selected intervals providing data to determine pressure gradients. This data can be used to define zcp levels and define reservoir compartments (Fig. 14).

The pressure in the water phase depends upon the degree to which the fluid column is connected to the Earth's surface. In an open system, the fluid pressure is equal to depth times the density of the fluid and is called hydrostatic (Fig. 15). The hydrostatic pressure gradient is about 0.434 psi/ft; overburden pressure equals the weight of the overburden sediment and has a gradient of about 1 psi/ft. Deviations from hydrostatic pressure, abnormal pressures, occur when the formation fluid is confined and cannot equilibrate with surface pressure. Overpressuring is the most common abnormal pressure and is produced by (1) compaction during rapid burial, (2) tectonic compression, and (3) hydrocarbon generation and migration (Osborne and Swarbrick 1997). In extreme cases, fluid pressures can equal and even exceed overburden pressures. Uncommonly, pressures can be lower than hydrostatic. Underpressure is often related to erosional unloading that results in an increase in pore volume due to the elastic rebound of the sediment as the overburden is reduced (Bachu and Underschultz 1995).

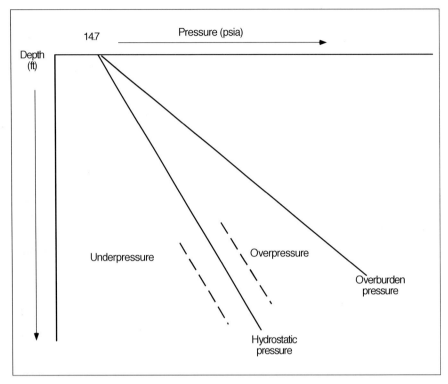

Fig. 15. Diagram illustrating overburden, normal hydrostatic, and abnormal over- and underpressure regimes. (After Dake 1978)

1.5
Relative Permeability

Oil, water, and gas are found in hydrocarbon reservoirs in varying proportions. Permeability measurements, however, are typically done using a single fluid, commonly air or water, and the permeability values must be corrected for the varying saturations of water, oil, and gas that occur in the reservoir. The correction is necessary because when a non-wetting fluid, such as oil, enters a water-wet pore system, the oil fills the centers of the largest, well-connected pores, whereas the water is found lining the pore walls and filling the smallest pores. This fluid distribution reduces the pore space available for flow of either water or oil. When water is injected or imbibed into the water-wet pore system, oil is trapped in pores with the smallest pore throats due to capillary forces. This oil is referred to as residual oil to water flooding (Fig. 16).

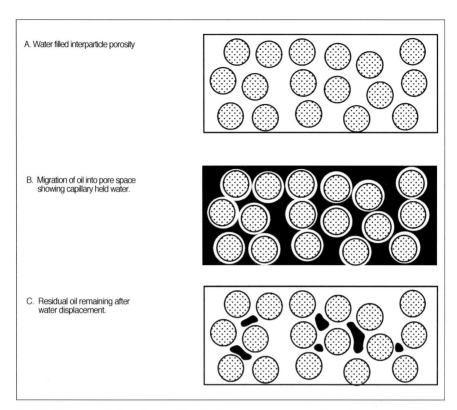

A. Water filled interparticle porosity

B. Migration of oil into pore space showing capillary held water.

C. Residual oil remaining after water displacement.

Fig. 16. Diagram of oil and water distribution in a water-wet rock under three conditions; A 100% water saturation, B injection of a nonwetting fluid (oil), and C injection of a wetting fluid (water)

Relative permeability is simply the permeability measured at a specific fluid saturation expressed as a fraction of the total or absolute permeability. Absolute permeability is the permeability of a rock that is 100% saturated with a single fluid. In a water-wet rock only water can totally saturate the pore system, and brine permeability is normally taken as the absolute permeability. However, hydrocarbon permeability at residual water saturation is often used as absolute permeability in reservoir engineering studies. Effective permeability is the permeability of one fluid in the presence of another fluid measured at a specific saturation state. Effective permeability is always lower than the absolute permeability and will change as the saturation changes. Thus, if a rock 100% saturated with brine has a permeability of 50 md whereas the brine permeability in the presence of 50% oil saturation is 10 md, the relative permeability of brine at 50% oil saturation is said to be 0.2. Graphs of relative permeability versus saturation (Fig. 17) are very important because they can be used to predict changes in production rates with changes in water satura-

Fig. 17. Typical relative permeability plot where absolute permeability is taken as oil permeability at lowest water saturation. Kro is the relative permeability to oil and Krw is the relative permeability to water

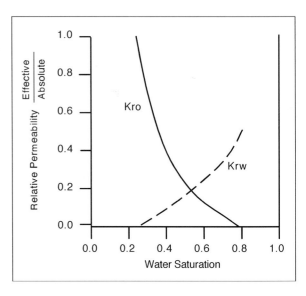

tion. They are fundamental in fluid flow simulation, and changing the relative permeability characteristics has a major effect on the resulting performance prediction.

There are two methods for measuring permeability at various saturation states to obtain relative permeability, steady state and unsteady state (Fig. 18). The steady-state method is the most accurate method but is time consuming and expensive because it involves injecting both oil and water simultaneously until the output rates match the input rates. The unsteady-state method is less accurate but faster because it involves saturating the core with oil and flooding with water. The relationships between relative permeability and saturation obtained by these two methods are commonly very different. A third method that is rapid and less expensive is to measure effective permeabilities at irreducible water and residual oil. This is called the end point method and assumes that a reasonable estimate of the curvature can be made.

A significant problem in measuring relative permeability in the laboratory is restoring samples to reservoir conditions. Pore surfaces, especially in carbonate rocks, are reactive to changes in fluids, and these reactions can alter the wettability state. Elaborate methods have been devised to preserve the original wettability state of core material, and the accuracy of any relative permeability data is dependent upon the success of these methods. Many carbonate reservoirs are considered to have mixed wettability at present; some pore walls are oil wet and some water wet. However, it is most likely that the reservoirs were water wet at the time of oil migration.

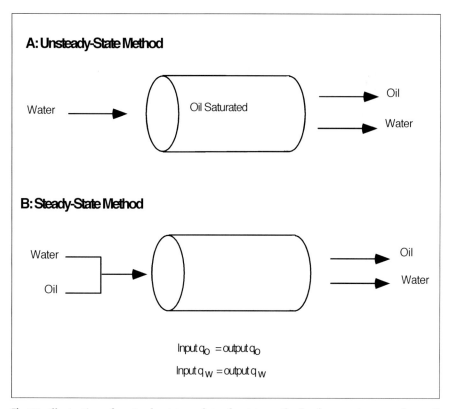

Fig. 18. Illustration of unsteady-state and steady-state methods of measuring two-phase oil and water relative permeability.

Reservoir height (capillary pressure), relative permeability, and saturation are interrelated as illustrated in Fig. 19 (Arps 1964). The oil becomes mobile only after attaining a saturation defined by the relative permeability curve that equates to a reservoir height defined by the capillary pressure curve. This level often defines the field oil/water contact. Oil and water are produced above this reservoir height until the relative permeability to water becomes extremely low and only oil will flow. The reservoir height at which this occurs is defined by the capillary pressure curve. This depth interval is commonly referred to as the transition zone between water production and oil production. Above this interval, water free oil production can be expected.

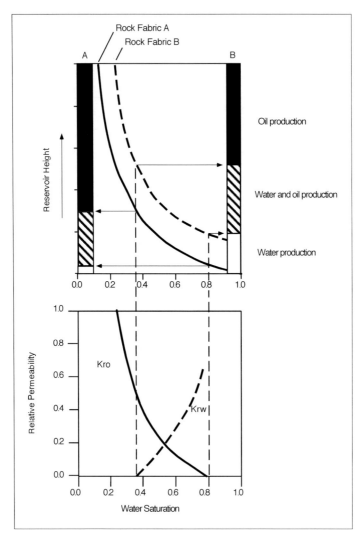

Fig. 19. Simplified illustration showing the relationship between relative permeability to oil and water, capillary pressure converted to reservoir height, and water saturation. The effect of rock fabric is illustrated by considering two capillary pressure curves (A, B) from carbonate rocks with different pore-size distributions. The change in pore size results in the possibility of intervals where (1) clean oil is produced from rock-fabric A and oil and water from rock-fabric B , and (2) oil and water is produced from rock-fabric A and water from B

1.6
Summary

The petrophysical properties of porosity, permeability, relative permeability, and capillarity are linked through pore-size distribution. Porosity is a fundamental property of reservoir rocks but is not related to pore size; pore-size distribution is related to the fabric of the rock of which porosity is one element. Fluid saturations, such as water saturation, are a function of pore-size distribution and reservoir height; reservoir height being equated with capillary pressure. Permeability is a function of pore-size distribution, interparticle porosity, and fluid saturations.

Petrophysical properties must be tied to geological descriptions in order to display these properties in realistic 3D images. The link is through pore-size distribution. Porosity and rock fabric together determine pore-size distribution. Porosity is measured using various visual and laboratory methods, the most accurate being the Boyles Law method. Rock fabric, however, is a interpretive geological description that can be tied to geologic models and sequence stratigraphy. Methods of obtaining this information are discussed in Chapter 2.

References

Amyx JW, Bass DM Jr, Whiting RL (1960) Petroleum reservoir engineering. McGraw-Hill, New York, 610 pp

Archie GE (1952) Classification of carbonate reservoir rocks and petrophysical considerations. AAPG Bull 36, 2: 278–298

Arps JJ (1964) Engineering concepts useful in oil finding. AAPG Bull 43, 2: 157–165

Bachu S, Underschultz JR (1995) Large-scale underpressuring in the Mississippian-Cretaceous succession, Southwestern Alberta Basin. AAPG Bull 79, 7: 989–1004

Beard DC, Weyl PK (1973) Influence of texture on porosity and permeability in unconsolidated sand. AAPG Bull 57: 349–369

Dake LP (1978) Fundamentals of reservoir engineering: developments in petroleum science, 8. Elsevier, Amsterdam, 443 pp

Enos P, Sawatsky LH (1981) Pore networks in Holocene carbonate sediments. J Sediment Petrol 51, 3: 961–985

Harari Z, Sang Shu-Tek, Saner S (1995) Pore-compressibility study of Arabian carbonate reservoir rocks. SPE Format Eval 10, 4: 207–214

Hurd BG, Fitch JL (1959) The effect of gypsum on core analysis results. J Pet Technol 216: 221–224

Hurst A, Goggin D (1995) Probe permeametry: an overview and bibliography. AAPG Bull 79, 3: 463–471

Kozeny JS (1927) (no title available). Ber Wiener Akad Abt Iia, 136: p 271

Lucia FJ (1995) Rock-Fabric/petrophysical classification of carbonate pore space for reservoir characterization. AAPG Bull 79, 9: 1275–1300

Osborne MJ, Swarbrick RE (1997) Mechanisms for generating overpressure in sedimentary basins: a reevaluation. AAPG Bull 81, 6: 1023–1041

Pittman ED (1992) Relationship of porosity and permeability to various parameters derived from mercury injection-capillary pressure curves for sandstone. AAPG Bull 72, 2: 191–198

Schmoker JW, Krystinic KB, Halley RB (1985) Selected characteristics of limestone and dolomite reservoirs in the United States. AAPG Bull 69, 5: 733–741

Smolen JJ, Litsey LR (1977) Formation evaluation using wireline formation tester pressure da-
    ta. SPE paper 6822, presented at SPE-AIME 1977 Fall Meeting, Oct 6–12, Denver, Colorado
Swanson BJ (1981) A simple correlation between permeability and mercury capillary pres-
    sures. J Pet Technol Dec: 2488–2504

CHAPTER 2

Rock-Fabric, Petrophysical Parameters, and Classification

2.1
Introduction

The goal of reservoir characterization is to describe the spatial distribution of petrophysical parameters such as porosity, permeability, and saturation. Wireline logs, core analyses, production data, pressure buildups, and tracer tests provide quantitative measurements of petrophysical parameters in the vicinity of the wellbore. This wellbore data must be integrated with a geologic model to display the petrophysical properties in three-dimensional space. Studies that relate rock fabric to pore-size distribution, and thus to petrophysical properties, are key to quantification of geologic models in numerical terms for input into computer simulators (Fig. 1).

Geologic models are generally based on observations that are interpreted in terms of depositional models and sequences. In the subsurface, cores, wireline

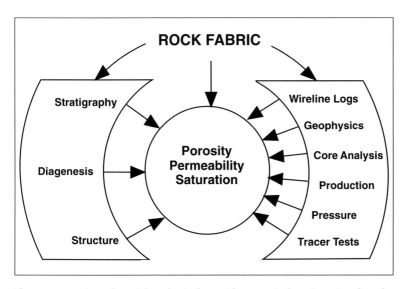

Fig. 1. Integration of spatial geologic data with numerical engineering data through rock-fabric studies

logs, and seismic data are the main sources of information for these interpretations. Engineering models are based on wireline log calculations and average rock properties from core analyses. Numerical engineering data and interpretive geologic data are joined at the rock fabric level because the pore structure is fundamental to petrophysical properties and the pore structure is the result of spatially distributed depositional and diagenetic processes.

The purpose of this chapter is to define important geologic parameters which when described and mapped allow accurate petrophysical quantification of carbonate geologic models by (1) describing the relationship between carbonate rock fabrics and petrophysical properties and (2) presenting a generic petrophysical classification of carbonate pore space.

2.2
Pore Space Terminology and Classification

Pore space must be defined and classified in terms of rock fabrics and petrophysical properties in order to integrate geological and engineering information. Archie (1952) made the first attempt at relating rock fabrics to petrophysical rock properties in carbonate rocks. The Archie classification focuses on estimating porosity but is also useful for approximating permeability and capillary properties. Archie (1952) recognized that not all the pore space can be observed using a X10 power microscope and that the surface texture of the broken rock reflected the amount of matrix porosity. Therefore, pore space is divided into matrix and visible porosity (Fig. 2). *Chalky texture* indicates a matrix porosity of about 15%, *sucrosic texture* indicates a matrix porosity of about 7%, and *compact texture* indicates matrix porosity of about 2%. Visible pore space is described according to pore size; A for no visible pore space and B,C, and D for increasing pore sizes from pinpoint to larger than cutting size. Porosity/permeability trends and capillary pressure characteristics are also related to these textures.

Although the Archie method is still useful for estimating petrophysical properties, relating these descriptions to geologic models is difficult because the descriptions cannot be defined in depositional or diagenetic terms. A principal difficulty is that no provision is made for distinguishing between visible interparticle pore space and other types of visible pore space such as moldic pores. Research on carbonate pore space (Murray 1960; Choquette and Pray 1970; Lucia 1983) has shown the importance of relating pore space to depositional and diagenetic fabrics and of distinguishing between interparticle (intergrain and intercrystal) and other types of pore space. Recognition of the importance of these factors prompted modification of Archie's classification.

The petrophysical classification of carbonate porosity presented by Lucia (1983, 1995) emphasizes petrophysical aspects of carbonate pore space, as does the Archie classification. However, by comparing rock fabric descriptions with laboratory measurements of porosity, permeability, capillarity, and Archie m values (Ro/Rw = Porosity $^{-m}$), Lucia (1983) showed that the most useful division of pore types was between pore space located between grains or crystals, called

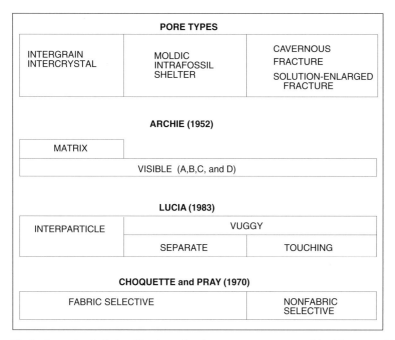

Fig. 2. Petrophysical classification of carbonate pore types used in this report (Lucia 1983) compared with Archie's original classification (1952) and the fabric selectivity concept of Choquette and Pray (1970)

interparticle porosity, and all other pore space, called vuggy porosity (Fig. 2). Vuggy pore space is further subdivided by Lucia (1983) into two groups based on how the vugs are interconnected; (1) vugs that are interconnected only through the interparticle pore network are termed *separate vugs* and (2) vugs that form an interconnected pore system are termed *touching vugs*.

Choquette and Pray (1970) discussed the geologic concepts surrounding carbonate pore space and presented a classification that is widely used. They emphasize the importance of pore space genesis, and the divisions in their classification are genetic and not petrophysical. They divide all carbonate pore space into two classes: fabric selective and nonfabric selective (Fig. 2). Moldic and intraparticle pore types are classified as fabric selective porosity by Choquette and Pray (1970) and grouped with interparticle and intercrystalline porosity. However, Lucia (1983) demonstrated that moldic and intraparticle pores have a different effect on petrophysical properties than do interparticle and intercrystalline pores and, thus should be grouped separately. Pore-type terms used in this classification are listed in Fig. 3 and compared with those suggested by Choquette and Pray. Although most of the terms defined by Choquette and Pray are also used here, interparticle and vug porosity have different definitions. Lucia (1983) demonstrated that pore space located both between grains (intergrain

| Term | Abbreviations | |
	Lucia	Choquette and Pray (1970)
Interparticle	IP	BP
Intergrain	IG	-
Intercrystal	IX	BC
Vug	VUG	VUG
Separate Vug	SV	-
Moldic	MO	MO
Intraparticle	WP	WP
Intragrain	WG	-
Intracrystal	WX	-
Intrafossil	WF	-
Intragrain microporosity	μG	-
Shelter	SH	SH
Touching Vug	TV	-
Fracture	FR	FR
Solution-enlarged fracture	SF	CH*
Cavernous	CV	CV
Breccia	BR	BR
Fenestral	FE	FE
*Channel.		

Fig. 3. Pore-type terminology used in this report compared with terminology of Choquette and Pray (1970)

porosity) and between crystals (intercrystal porosity) are petrophysically similar and a term is needed to identify these petrophysically similar pore types. The term "interparticle" was selected because of its broad connotation. The classification of Choquette and Pray (1970) does not have a term that encompasses these two petrophysically similar pore types. In their classification, the term "interparticle" is used instead of "intergrain".

Vuggy porosity, as defined by Lucia (1983), is pore space that is within grains or crystals or that is significantly larger than grains or crystals; that is pore space that is not interparticle. Vugs are commonly present as dissolved grains, fossil chambers, fractures, and large irregular cavities. Although fractures may not be formed by depositional or diagenetic processes, fracture porosity is included because it defines a unique type of porosity in carbonate reservoir rocks. This definition of vug deviates from the restrictive definition of vugs used by Choquette and Pray (1970) as nondescript, nonfabric selective pores, but it is consistent with the Archie terminology and with the widespread and less restrictive use in the oil industry of the term "vuggy porosity" to refer to visible pore space in carbonate rocks.

2.3
Rock-Fabric/Petrophysical Classification

The foundation of the Lucia classification, as well as the Archie classification, is the concept that pore-size distribution controls permeability and saturation and that pore-size distribution is related to rock fabric. In order to relate carbonate rock fabrics to pore-size distribution, it is important to determine if the pore space belongs to one of the three major pore-type classes, interparticle, separate-vug, or touching-vug. Each class has a different type of pore-size distribution and interconnection. It is equally important to determine the volume of pore space in these various classes because pore volume relates to reservoir volume and, in the case of interparticle and separate-vug porosity, to pore-size distribution.

2.3.1
Classification of Interparticle Pore Space

In the absence of vuggy porosity, pore-size distribution in carbonate rocks can be described in terms of particle size, sorting and interparticle porosity (Fig. 4).

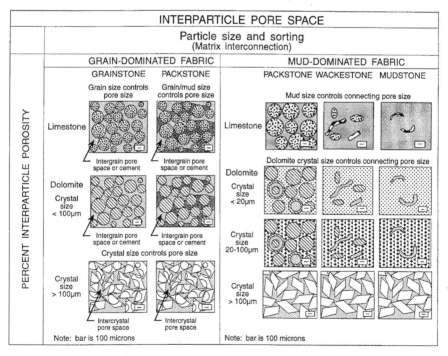

Fig. 4. Geological/petrophysical classification of carbonate interparticle pore space based on size and sorting of grains and crystals. The volume of interparticle pore space is important because it relates to pore-size distribution

Fig. 5. Relationship between mercury displacement pressure and average particle size for nonvuggy carbonate rocks with permeability greater than 0.1 md (Lucia 1983). The displacement pressure is determined by extrapolating the capillary pressure curve to a mercury saturation of zero

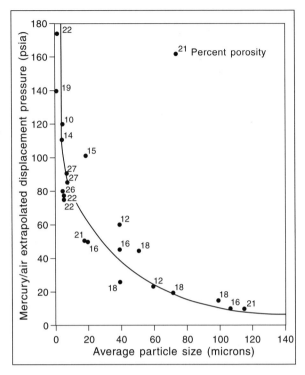

Lucia (1983) showed that particle size can be related to mercury capillary displacement pressure in nonvuggy carbonates with more the 0.1 md permeability, suggesting that particle size describes the size of the largest pores (Fig. 5). Whereas the displacement pressure characterizes the largest pore sizes and is largely independent of porosity, the shape of the capillary pressure curve characterizes the smaller pore sizes and is dependent on interparticle porosity (Lucia 1983).

The relationship between displacement pressure and particle size (Fig. 5) is hyperbolic and suggests important particle-size boundaries at 100 and 20 μm. Lucia (1983) demonstrated that three permeability fields can be defined using particle-size boundaries of 100 and 20 μm, a relationship that appears to be limited to particle sizes of less than 500 μm (Fig. 6).

Recent work has shown that permeability fields can be better described in geologic terms if sorting as well as particle size is considered. The approach to size and sorting used in this petrophysical classification is similar to the grain-/mud-support principle upon which the Dunham's (1962) classification is built. Dunham's classification, however, is focused on depositional texture whereas petrophysical classifications are focused on contemporary rock fabrics which include depositional and diagenetic textures. Therefore, minor modifications must be made in Dunham's classification before it can be applied to a petrophysical classification.

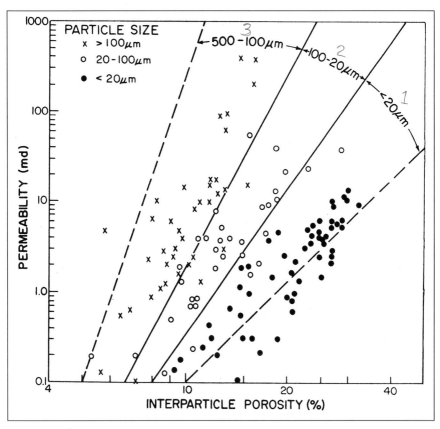

Fig. 6. Porosity-air permeability relationship for various particle-size groups in nonvuggy carbonate rocks (Lucia 1983)

Instead of dividing fabrics into grain support and mud support as in Dunham's classification, fabrics are divided into grain-dominated and mud-dominated (Fig. 4). The important attributes of grain-dominated fabrics are the presence of open or occluded intergrain porosity and a grain-supported texture. The important attribute of mud-dominated fabrics is that the areas between the grains are filled with mud even if the grains appear to form a supporting framework.

Grainstone is clearly a grain-dominated fabric, but Dunham's packstone class bridges a boundary between large intergrain pores in grainstone and small interparticle pores in wackestones and mudstones. Some packstones have intergrain pore space and some have the intergrain spaces filled with mud. The packstone textural class must be divided into two rock-fabric classes: grain-dominated packstones that have intergrain pore space or cement and mud-dominated packstones that have intergrain spaces filled with mud (Fig. 4).

2.3.2
Classification of Vuggy Pore Space

The addition of vuggy pore space to interparticle pore space alters the petro-physical characteristics by altering the manner in which the pore space is con-nected, all pore space being connected in some fashion. Separate vugs are de-fined as pore space that is interconnected only through the interparticle porosi-ty. Touching vugs are defined as pore space that forms an interconnected pore system independent of the interparticle porosity (Fig. 7).

2.3.2.1
Separate-Vug Pore Space

Separate-vug pore space is defined as pore space that is 1) either within particles or is significantly larger than the particle size (generally >2 x particle size), and 2) is interconnected only through the interparticle porosity (Fig. 7). Separate vugs are typically fabric-selective in their origin. Intrafossil pore space, such as the liv-ing chambers of a gastropod shell; moldic pore space, such as dissolved grains (oomolds) or dolomite crystals (dolomolds); and intragrain microporosity are ex-amples of intraparticle, fabric-selective separate vugs. Molds of evaporite crystals and fossil molds found in mud-dominate fabrics are examples of fabric-selective separate vugs that are significantly larger than the particle size. In mud-dominat-ed fabrics, shelter pore space is typically much larger than the particle size and is classified as separate-vug porosity whereas in grain-dominated fabrics, shelter pore space is related to particle size and is considered intergrain porosity.

In grain-dominated fabrics, extensive selective dissolution of grains may cause grain boundaries to dissolve, producing composite molds. These composite molds may have the petrophysical characteristics of separate vugs. However, if dissolution of the grain boundaries is extensive, the pore space may be intercon-nected well enough to be classified as solution-enlarged interparticle porosity.

Grain-dominated fabrics may contain grains with intragrain microporosity (Pittman 1971). Even though the pore size is small, intragrain microporosity is classified as a type of separate vug because it is located within the particles of the rock. Mud-dominated fabrics may also contain grains with microporosity, but they present no unique petrophysical condition because of the similar pore sizes between the microporosity in the mud matrix and in the grains.

2.3.2.2
Touching-Vug Pore Space

Touching-vug pore systems are defined as pore space that is (1) significantly larger than the particle size and (2) forms an interconnected pore system of sig-nificant extent (Fig. 7). Touching vugs are typically nonfabric selective in origin. Cavernous, breccia, fracture, and solution-enlarged fracture pore types com-monly form an interconnected pore system on a reservoir scale and are typical

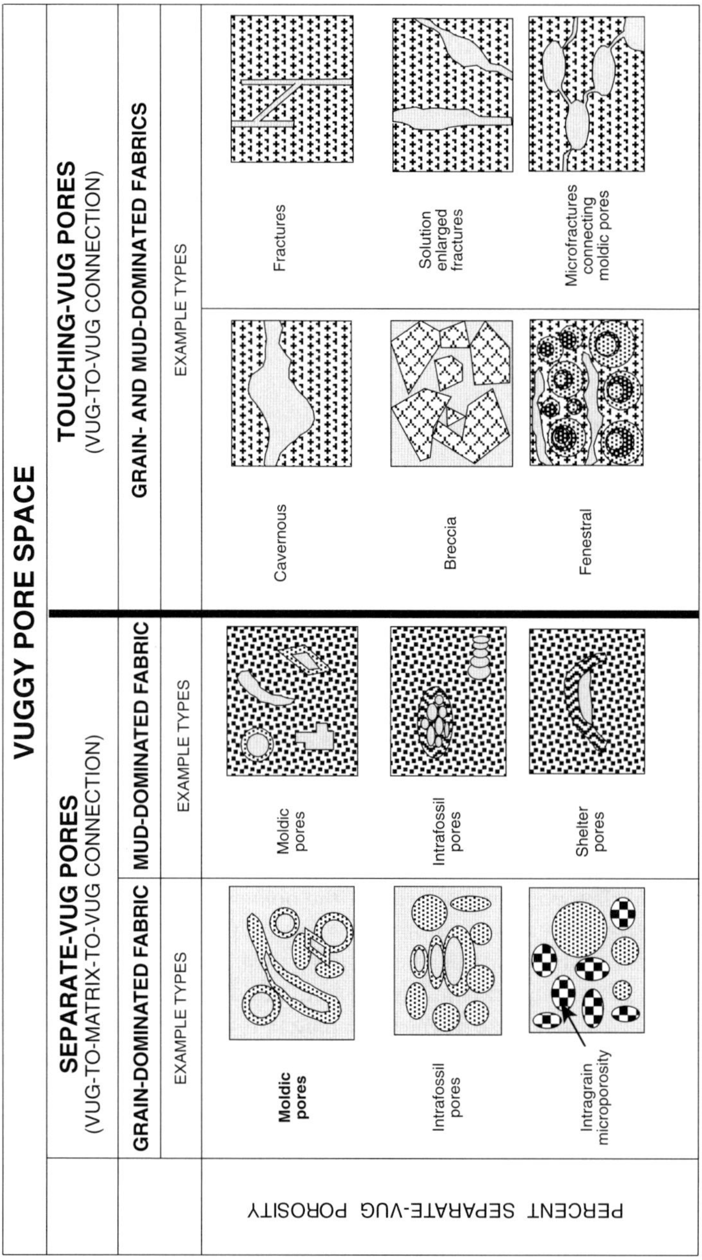

Fig. 7. Geological/petrophysical classification of vuggy pore space based on vug interconnection. The volume of separate-vug pore space is important for characterizing the pore-size distribution

touching-vug pore types. Fenestral pore space is commonly connected on a reservoir scale and is grouped with touching vugs because the pores are normally much larger than the grain size (Major et al. 1990).

Fracture porosity is included as a touching-vug pore type because fracture porosity is an important contributor to permeability in many carbonate reservoirs and, therefore, must be included in any petrophysical classification of pore space. Although fracturing is often considered to be of tectonic origin and thus not a part of carbonate geology, diagenetic processes common to carbonate reservoirs, such as karsting (Kerans 1989), can produce extensive fracture porosity. The focus of this classification is on petrophysical properties rather than genesis, and must include fracture porosity as a pore type irrespective of its origin.

2.4
Rock-Fabric/Petrophysical Relationships

2.4.1
Interparticle Porosity/Permeability Relationships

2.4.1.1
Limestone Rock Fabrics

Examples of nonvuggy limestone petrophysical rock fabrics are illustrated in Fig. 8. In grainstone fabrics, the pore-size distribution is controlled by grain size; in mud-dominated fabrics, the size of the micrite particles controls the pore-size distribution. In grain-dominated packstones, however, the pore size distribution is controlled by grain size and by the size of micrite particles between grains. Fig. 9a illustrates a plot of air permeability against interpartical porosity for grainstones. The data is from Choquette and Steiner's (1985) publication on the Ste. Genevieve oolite (Mississippian). The average grain size of the oolites is about 400 µm. The points on this graph are concentrated within the >100 µm permeability field.

Figure 9c illustrates a plot of air permeability against interparticle porosity from microporous wackestones and mudstones from Middle Eastern Cretaceous reservoirs. The average crystal size of the mud matrix is about 5 µm. The data are concentrated in the <20 µm permeability field and define the lower limit of this field.

Because grain-dominated packstone is a new fabric class, data are difficult to find. Lucia and Conti (1987) report on a nonvuggy grain-dominated packstone of Wolfcampian age that occurs in a core taken in Oldham County, Texas Panhandle. The grain-dominated packstone is described as a poorly sorted mixture of 150 to 300 µm grains in a matrix composed of 80 µm pellets and 10 µm calcite crystals, clearly a poorly sorted texture. A porosity permeability plot of this data (Fig. 9b) shows that it plots on the boundary between 100- to 20-µm and the <20-µm permeability fields. An ooid-oncoid grain-dominated packstone has been described from Cretaceous reservoirs in the Santos Basin, offshore Brazil by Wagner Cruz (pers comm). The ooids are 400 µm in diameter, the oncoids 1–2 mm in di-

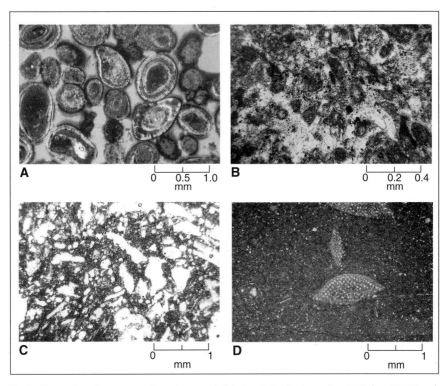

Fig. 8. Examples of nonvuggy limestone rock fabrics. A Grainstone, Ø = 25%, k = 15,000 md.
B Grain-dominated packstone, Ø = 16%, k = 5.2 md. Note intergrain cement and pore
space. C Mud-dominated packstone, Ø = 18%, k = 4 md. Note microporosity. D Wackestone
Ø = 33%, k = 9 md

ameter, and the intergrain lime mud is composed of 5 µm particles. Data from
this fabric scatter about the upper limit of the 100-µm to 20-µm permeability
field. An additional few data points have been gleaned from the literature and
they plot in the 100- to 20-µm permeability field.

Fig. 9d illustrates a plot between air permeability and porosity for North Sea
coccolith chalk (Scholle 1977). The average size of the coccoliths is about 1 µm.
The data points plot below the <20-µm permeability field. The presence of intrafossil pore space in the coccolith grains probably accounts for the lower than expected permeability values in the high porosity ranges.

Figure 10 illustrates all the data for limestones compared with the permeability fields. Grainstone and mud-dominated fabrics are reasonably well constrained permeability fields. Although grain-dominated packstone fabrics plot
at an intermediate location between grainstones and mud-dominated limestones, they show more variability because of the large grain size difference between the two fabrics illustrated.

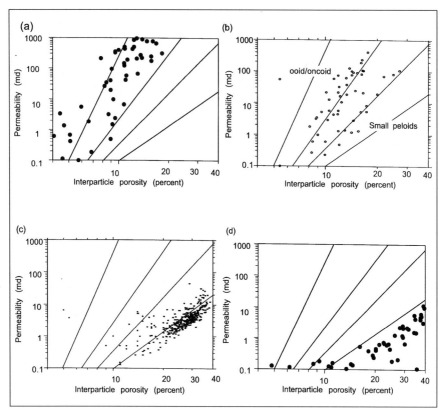

Fig. 9. Porosity-air permeability cross plots for nonvuggy limestone rock fabrics compared with the three permeability fields illustrated in Fig. 6. **A** 400-μm ooid grainstone, Ste. Genevieve, Illinois (Choquette and Steiner 1985). Low-permeability, high-porosity data are deleted because they are from oomoldic and wackestone rock fabrics (pers. comm from Choquette 1993). **B** Grain-dominated packstone data, Wolfcamp, West Texas (Lucia and Conti 1987). A poorly sorted mixture of 80- to 300-μm grains and micrite. Large ooids and oncoids, Cretaceous, Brazil. **C** Wackestones with microporosity between 5-μm crystals, Shuaiba, UAE (Moshier et al. 1988) and Qatar. Data associated with stylolites not shown. **D** Coccolith chalk, Cretaceous, (Scholle 1977). The presence of intragranular pore space in the coccoliths causes the data to plot below the <20-μm field

Despite the considerable scatter in the data, grainstone, grain-dominated packstone, and mud-dominated fabrics are reasonably well constrained to the three permeability fields. Whereas grain size and sorting define the permeability fields, the interparticle porosity defines pore-size distribution and thus the permeability within the field. Systematic changes in intergrain porosity by cementation, compaction, and dissolution processes will produce systematic changes in pore-size distribution and result in systematic changes in permeability. Therefore, permeability in nonvuggy limestones is a function of interparticle porosity, grain size, and sorting.

Fig. 10. Composite porosity-air permeability cross plot for nonvuggy limestone fabrics compared with the three permeability fields illustrated in Fig. 6. Chalks are not included because of the presence of intragrain pore space

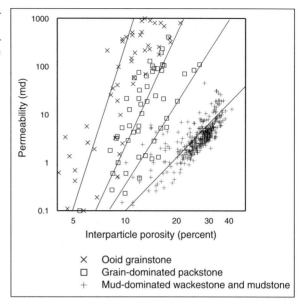

2.4.1.2
Dolomite Rock Fabrics

Examples of nonvuggy dolostone petrophysical rock fabrics are illustrated in Fig. 11. Dolomitization can change the rock fabric significantly. In limestones, fabrics can usually be distinguished with little difficulty. If the rock has been dolomitized, however, the overprint of dolomite crystals often obscures the limestone fabric precursor. Precursor fabrics in fine-crystalline dolostones are easily recognizable. However, as the crystal size increases, the precursor fabrics become progressively more difficult to determine.

Dolomite crystals (defined as particles in this classification) commonly range in size from several μm to >200 μm. Micrite particles are usually <20 μm in size. Thus, dolomitization of a mud-dominated carbonate fabric can result in an increase in particle size from <20 μm to >200 μm (Fig. 11). The plot of interparticle porosity against permeability (Fig. 12a) illustrates the principle that, in mud-dominated fabrics, permeability increases as dolomite crystal size increases. Finely crystalline (average 15 μm) mud-dominated dolostones from Farmer and Taylor Link (Lucia et al. 1992b) fields in the Permian Basin and from Choquette and Steiner (1985) plot within the <20 μm permeability field. Medium crystalline (average 50 μm) mud-dominated dolostones from the Dune field, Permian Basin, (Bebout et al. 1987), plot within the 100- to 20-μm permeability field. Large crystalline (average 150 μm) mud-dominated dolostones from Andrews

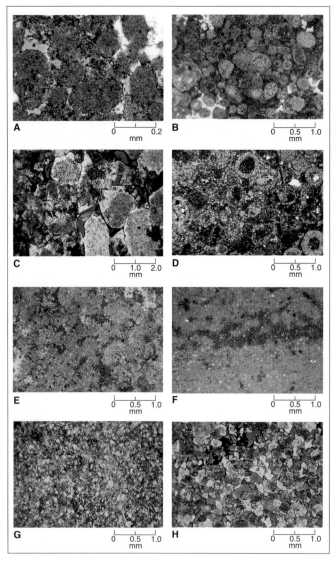

Fig.11. Examples of nonvuggy dolomite fabrics. **A** Dolograinstone, 15-μm dolomite crystal size, Ø = 16.4%, k = 343 md, Dune field (Bebout et al. 1987). **B** Dolograinstone, 30-μm dolomite crystal size, Ø = 7.1%, k = 7.3 md, Seminole San Andres Unit, West Texas. **C** Dolograinstone, crystal size 400 μm, Ø = 10.2%, k = 63 md, Harmatton field, Alberta, Canada. **D** Grain-dominated dolopackstone, 10-mm dolomite crystal size, Ø = 9%, k = 1 md, Farmer field, West Texas. **E** Grain-dominated dolopackstone, 30-μm dolomite crystal size, Ø = 9.5% k = 1.9 md, Seminole San Andres Unit, West Texas. **F** Fine crystalline dolowackestone, 10-mm dolomite crystal size, Ø = 11%, k = 0.12 md, Devonian, North Dakota. **G** Medium crystalline dolowackestone, 80-μm dolomite crystal size, Ø =16%, k = 30 md, Devonian, North Dakota. **H** Large crystalline dolowackestone, 150-μm dolomite crystal size, Ø = 20%, k = 4000 md, Andrews South Devonian field, West Texas

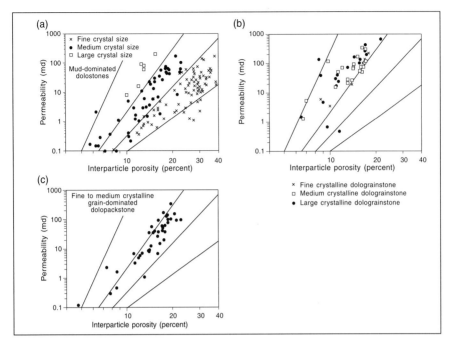

Fig. 12. Porosity-air permeability cross plots of nonvuggy dolomite fabrics compared with the three permeability fields illustrated in Fig. 6. **A** Mud-dominated dolostones with dolomite crystal sizes ranging from 10 to 150 μm. **B** Dolograinstones (average grain size is 200 mm) with dolomite crystal sizes ranging from 15 to 150 μm. **C** Grain-dominated dolopackstones with fine to medium dolomite crystal sizes

South Devonian field, Permian Basin (Lucia 1962), plot in the >100-μm permeability field.

Grainstones are usually composed of grains much larger than the dolomite crystal size (Fig. 11) so that dolomitization does not have a significant effect on the pore size distribution. This principle is illustrated in Fig. 12b where interparticle porosity is plotted against permeability measurements from dolomitized grainstones. The grain size of the dolograinstones is 200 μm. The finely crystalline dolograinstone from Taylor Link field, Permian Basin, the medium crystalline dolograinstone from Dune field, Permian Basin, and the large crystalline dolograinstone from an outcrop on the Algerita Escarpment, New Mexico, all plot within the >100-μm permeability field. The large crystalline mud-dominated dolostones also plot in this permeability field, indicating that they are petrophysically similar to grainstones (Fig. 12a).

Interparticle porosity and permeability measurements from fine to medium crystalline grain-dominated dolopackstones are crossplotted in Fig. 12c. The average grain size is 200 μm. The samples are from Seminole San Andres Unit and Dune (Grayburg) field (Bebout et al. 1987), Permian Basin. The data plot in the

Fig. 13. Composite porosity-air permeability cross plot for nonvuggy dolostone fabrics compared with the three permeability fields illustrated in Fig. 6.

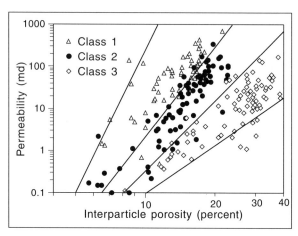

100- to 20-μm permeability field. The medium crystalline mud-dominated dolostones also plot in this field (Fig. 12a).

Fig. 13 illustrates all dolomite data compared with permeability fields. Dolograinstones and large crystal dolostones constitute the >100-μm permeability field. Grains are very difficult to recognize in dolostones with a >100 μm crystal size. However, because all large crystalline dolostones and all dolograinstones are petrophysically similar, whether the crystal size or the grain size is described makes little difference petrophysically. Fine and medium crystalline grain-dominated dolopackstones and medium crystalline mud-dominated dolostones constitute the 100- to 20-μm permeability field. Fine crystalline mud-dominated dolostones constitute the <20-μm field.

The dolomite permeability fields are defined by dolomite crystal size as well as grain size and sorting of the precursor limestone. Within the field, permeability is defined by interparticle porosity. Systematic changes in intergrain and intercrystal porosity by predolomite calcite cementation, dolomite cementation, and compaction will systematically change the pore-size distribution, resulting in a systematic change in permeability. Therefore, dolomite crystal size, grain size, and sorting define the permeability field, and interparticle porosity defines the permeability.

2.4.1.3
Limestone and Dolomite Comparison

Data from limestone and dolomite rock fabrics are combined into one porosity permeability cross plot in Fig. 14. The permeability fields are referred to as rock-fabric petrophysical Class 1, Class 2, and Class 3. The fabrics that make up the Class 1 field (>100-μm permeability field) are (1) limestone and dolomitized grainstones and (2) large crystalline grain-dominated dolopackstones and mud-dominated dolostones. The effect of grain size in this field can be seen by comparing Figs. 9a and 12b. Ooid grainstones, which have a grain size of 400 μm, are more permeable for a given porosity than dolograinstones, which have a grain size of 200 μm. The upper grain size limit of 500 μm is not well defined. An upper limit to this permeability field is imposed because as the grain size increases the slope of the porosity-permeability transform approaches infinity and porosity has little relationship to permeability.

Fabrics that make up the Class 2 field (100- to 20-μm permeability field) are (1) grain-dominated packstones, (2) fine to medium crystalline grain-dominated dolopackstones, and (3) medium crystalline mud-dominated dolostones. Grain size of the grain-dominated packstones and dolopackstones ranges from 400 microns of the oncoid/ooid packstone, 200 μm for the pellet dolopackstone, and 80–150 μm for the fine-grained packstone.

The Class 3 field (<20-μm permeability field) is characterized by mud-dominated fabrics (mud-dominated packstone, wackestone, and mudstone) and fine crystalline mud-dominated dolostones. Although not apparent from the data

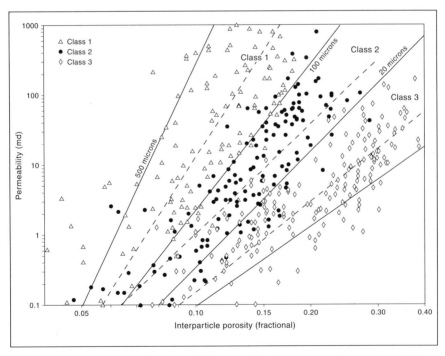

Fig. 14. Composite porosity-air permeability cross plot for nonvuggy limestones and dolostones showing statistical reduced-major-axis transforms for each class (*dashed lines*). See text for equations

presented, thin section observations suggest that permeability increases as the grain content increases within this field.

Reduced major axis (RMA) transforms are presented below for each combined permeability field (Fig. 14). The transform for the Class 2 field is slightly skewed to the field boundaries and a transform that is more compatible with the field boundaries is presented and recommended:

Class 1 $k = (45.35 \times 10^8) \times \Phi_{ip}^{8.537}$ $\qquad\qquad$ $r = 0.71$,

Class 2 $k = (1.595 \times 10^5) \times \Phi_{ip}^{5.184}$ $\qquad\qquad$ $r = 0.80$

$\qquad$ [recommended Class 2 $k = (2.040 \times 10^6) \times \Phi_{ip}^{6.38}$],

Class 3 $k = (2.884 \times 10^3) \times \Phi_{ip}^{4.275}$ $\qquad\qquad$ $r = 0.81$,

where k is in md, and Φ_{ip} is the fractional interparticle porosity.

Although fabrics are divided into three petrophysical classes, in nature there is no boundary between the classes. Instead, there is a continuum from mudstone to grainstone and from 5 μm mud-dominated dolostones to over 500 μm mud-dominated dolostones (Fig. 15a,b). Therefore, there is also a complete continuum of rock-fabric specific porosity-permeability transforms.

To model such a continuum the boundaries of each petrophysical class are assigned a value (0.5, 1.5, 2.5, and 4) (Fig. 15c) and porosity-permeability transforms generated. An equation relating permeability to a continuum of petrophysical classes and interparticle porosity is developed using multiple linear regressions. The resulting global transform is given below.

$$\text{Log}(k) = (A + B*\log(\text{class})) + (C + D*\text{Log}(\text{class}))(\Phi ip),$$

where A = 9.7982, B = 12.0838, C = 8.6711, D = 8.2065, class is the rock fabric petrophysical class ranging from 0.5 - 4, and Φip is the fractional interparticle porosity.

Mud-dominated limestones and fine crystalline mud-dominated dolostones occupy classes from 4 to 2.5 (Fig. 15a,b). The class number decreases with increasing dolomite crystal size from 5 to 20 μm in mud-dominated dolostones, and with increasing grain volume in mud-dominated limestones. Grain-dominated packstones, fine-to-medium crystalline grain-dominated dolopackstones, and medium crystalline mud-dominated dolostones occupy the classes from 2.5 to 1.5 (Fig. 15a,b). The class value decreases with increasing dolomite crystal size from 20 to 100 μm in mud-dominated dolostones and with decreasing amounts of intergrain micrite as well as increasing grain size in grain-dominated packstones and fine to medium crystalline grain-dominated dolopackstones. Grainstones, dolograinstones, and large crystalline dolomites occupy classes 1.5 to 0.5 (Fig. 15a,b). The class value decreases with increasing grain size and dolomite crystal size from 100 to 500 μm.

Fig. 15. Continuum of rock fabrics and associated porosity-permeability transforms. **A** Class fabrics ranging from 0.5 – 4 defined by class-average and class-boundary porosity-permeability transforms. **B** Fabric continuum in nonvuggy limestone. **C** Fabric continuum in nonvuggy dolostone

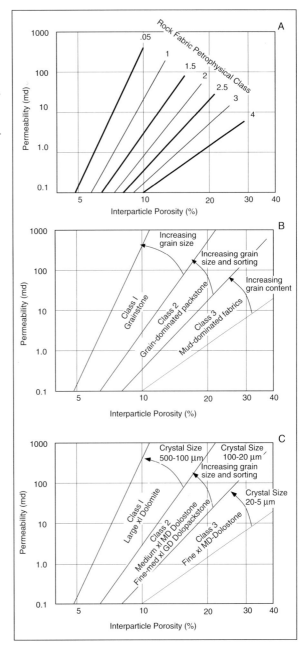

2.4.1.4
Unusual Types of Interparticle Porosity

Diagenesis can produce unique types of interparticle porosity. Collapse of separate-vug fabrics due to overburden pressure can produce fragments that are properly considered "diagenetic particles". Large dolomite crystals with their centers dissolved can collapse to form pockets of dolomite rims. Leached grainstones can collapse to form intergrain fabrics composed of fragments of dissolved grains. These unusual pore types typically do not cover an extensive area. However, the collapse of extensive cavern systems can produce bodies of collapse breccia that are extensive. Interbreccia-block pores produced by cavern collapse are included in the touching-vug category because they result from the karsting process (Kerans 1989).

2.4.2
Interparticle Porosity/Water Saturation Relationships

Several methods have been presented for relating porosity, permeability, water saturation and reservoir height (Leverett 1941; Aufricht and Koepf 1957; Heseldin 1974; Alger et al. 1989). These methods attempt to average the capillary pressure curves into one relationship between saturation, porosity, permeability and reservoir height. The Leverett "J" function is a common method of averaging capillary pressure data. A common form of the J function is given below.

$$J(Sw) = (Pc/\sigma)(k/\phi)^{1/2} ,$$

Where Pc is capillary pressure, σ is interfacial tension, K is permeability in md, and ϕ is the fractional porosity.

The Leverett J function relates water saturation to capillary pressure, which is a function of reservoir height, and $(k/\phi)^{1/2}$, which is a function of pore size. As will be discussed later in this chapter, pore size can be related to rock fabric and interparticle porosity as well as to $(k/\phi)^{1/2}$. Therefore, changes in interparticle porosity within each rock-fabric field represents a change in pore size, and water saturation should be related to reservoir height, interparticle porosity, and rock fabric:

$$Sw = f(Pc, \phi_{ip}, \text{rock-fabric}).$$

To quantify the saturation characteristics of the three rock-fabric fields, capillary pressure curves with different interparticle porosities from each rock-fabric field are compared (Fig. 16). Curves from two samples of fine to medium crystalline dolograinstone representative of class 1 are presented in Fig. 16a. The 9.23% porosity curve represents the average of two sets of capillary pressure data from dolograinstones in the Taylor Link field (Lucia et al. 1992b) and the 17.6% porosity curve represents the average of two data sets, one from the Taylor Link field and one from the Dune field (Bebout et al. 1987). Curves from three samples

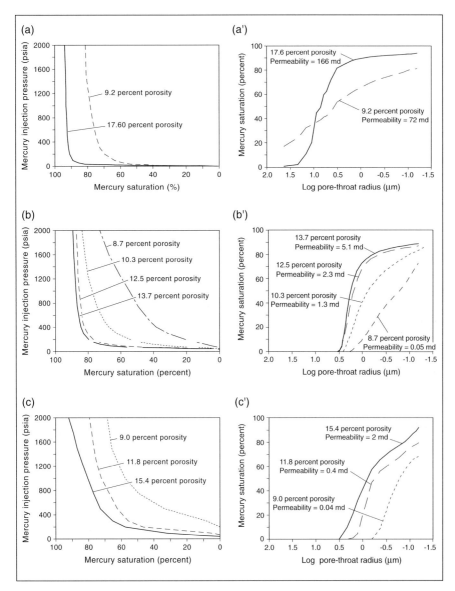

Fig. 16. Capillary pressure curves and pore-size distribution for petrophysical classes. **A** Class 1. Data presented are from dolograinstones. **B** Class 2. Data presented are from medium crystalline dolowackestones. **C** Class 3. Data are from fine crystalline dolowackestones

Table 1. Values used to convert mercury/air capillary
pressure to reservoir height

	Laboratory Mercury/air/solid	Reservoir Oil/water/solid
σ(dynes/cm)	480	28
θ (degrees)	140	44
Water density (gr/cc)	1.04	0.88

of fine crystalline dolowackestones representative of class 3 are presented in Fig. 16c. They are from the Farmer field, Permian Basin. Three curves representative of class 2 are presented in Fig. 16b. Class 2 represents a very diverse class of rock fabrics, and it is difficult to combine all the fabrics into a few simple curves. The curves presented here are from medium crystalline dolowackestones of the Seminole San Andres Unit and may not be representative of all grain-dominated packstones.

Each group of curves is characterized by similar displacement pressures and a systematic change in curve shape and saturation characteristics with changes in interparticle porosity. Equations relating water saturation to porosity and reservoir height are developed using a multiple linear regression with the log of water saturation as the dependent variable and the logs of capillary pressure and porosity as independent variables. Mercury capillary pressure is converted to reservoir height using generic values (Table 1).

The resulting equations are listed below and three dimensional representations for classes 1 and 3 are presented in Fig. 17.

Class 1: $Sw = 0.02219 \times H^{-0.316} \times \Phi_{ip}^{-1.745}$

Class 2: $Sw = 0.1404 \times H^{-0.407} \times \Phi_{ip}^{-1.440}$

Class 3: $Sw = 0.6110 \times H^{-0.505} \times \Phi_{ip}^{-1.210}$

The relationship between porosity, saturation and rock fabric class can be demonstrated by selecting a reservoir height of 500 ft (equates to a mercury capillary pressure of about 650 psia) and plotting saturation against porosity for each rock-fabric class. The results (Fig. 18) show that in nonvuggy carbonate reservoirs, a plot of porosity versus water saturation can separate the three rock-fabric groups into saturation fields similar to permeability fields.

The rock-fabric-specific saturation equations are specific to the reservoir conditions given in Table 1. However, the equations will provide reasonable values for original water saturations when porosity and rock fabric data are all that is available. This becomes important in old fields where resistivity logs are not available and where waterflooding has occurred. Intervals flooded by injection water can be identified by comparing the original water saturation values estimated by the above equations with current water saturations calculated from porosity and resistivity logs. An example from the Seminole San Andres field,

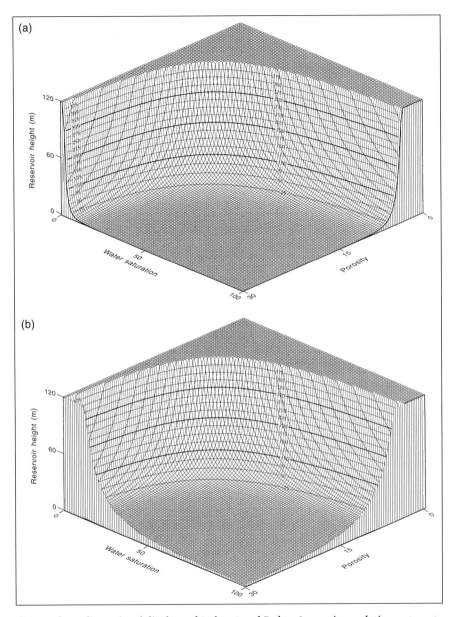

Fig. 17. Three-dimensional displays of A class 1 and B class 3 equations relating water saturation to reservoir height and porosity

Fig. 18. Cross plot of water saturation and porosity for the three rock-fabric petrophysical classes at a reservoir height of 150 m (500 ft). Water saturation (1-Hg saturation) and porosity values are taken from capillary pressure curves illustrated in Fig. 16

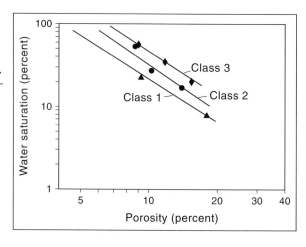

Fig. 19. Example of rock fabric/capillary pressure method of estimating original water saturation in water flooded wells, and identifying intervals invaded by injected water. Well 2709 was free of water at completion whereas well 2714 produced considerable flood water at completion. Correlation of rock fabrics between 2709 and 2714 provides the information needed to use rock-fabric specific saturation-porosity-height equations to estimate prewaterflood water saturation. A comparison of log calculated saturations with prewaterflood saturations indicates intervals invaded by injection water

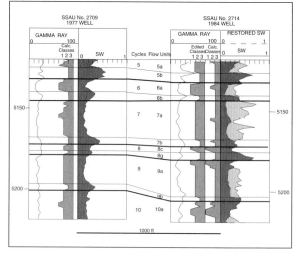

free completion) is compared with a partially flooded well (based on a high water cut completion). The rock fabrics are based on interpolation from cored wells using a high-frequency cycle correlation, and Rw is based on produced waters. Flooded intervals are identified by (1) a higher calculated than estimated Sw and (2) a high calculated Sw in the more porous intervals than in the less porous intervals.

Pore-throat-size distribution for each capillary pressure curve was calculated using the following formula.

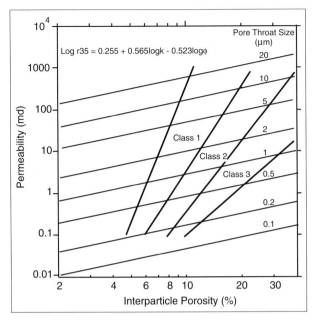

Fig. 20. Comparison of petrophysical classes and pore size versus interparticle porosity and permeability (After Pittman, 1992)

$$Ri = (2\sigma \times \cos\theta \times C)/Pc,$$

where Ri is the pore-throat radius (μm), σ = air-mercury interfacial tension (480 dynes/cm), θ is the air-mercury contact angle (140°), C is the unit conversion constant (0.145), and Pc is the mercury injection pressure (psia). The results (Fig. 16) show decreasing pore-throat size with decreasing porosity within a petrophysical class and a general decrease in average pore-throat size from class 1 to class 3. Microporosity (pore-throat size less than 1 μm (Pittman 1971)) is concentrated in class 3 and decreases in importance from class 3 to class 1.

Pittman (1992) has published relationships between interparticle porosity, permeability, and capillary pressure for siliciclastics. He concludes that pore-throat size measured at 35% mercury saturation gives the best relationship to porosity and permeability, and his equation is:

$$Log_{r35} = 0.255 + 0.565 \log k - 0.523 \log \Phi$$

This petrophysical approach is difficult to relate to geologic descriptions because rock fabrics do not have unique pore-throat sizes. The above equation is plotted in Fig. 20 and compared with the petrophysical class fields described in this report. From Fig. 20 it is apparent that Pittman's relationship between pore-throat sizes, porosity, and permeability does not conform to the rock-fabric classes defined in this classification, and therefore cannot be used to quantify geologic descriptions in terms of permeability. It is also apparent from Fig. 20 that pore

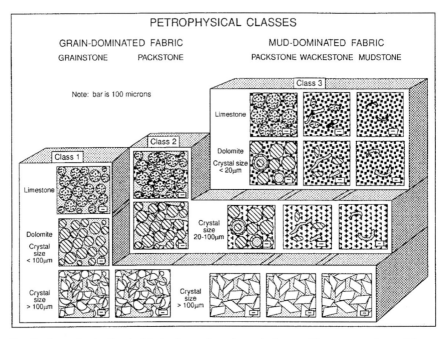

Fig. 21. Petrophysical/rock-fabric classes based on similar capillary properties and interparticle porosity/permeability transforms

throat size is related to rock-fabric class and interparticle porosity, and that within a fabric class, pore-throat size decreases as interparticle porosity decreases.

2.4.3
Rock-Fabric/Petrophysical Classes

Because the three rock-fabric groups define permeability and saturation fields, they, together with interparticle porosity and reservoir height, can be used to relate petrophysical properties to geologic observations. These rock-fabric groups, herein termed rock-fabric/petrophysical classes (Fig. 21) are described with their generic transform equations as follows:

Class 1 is composed of grainstones, dolograinstones, and large crystalline dolostones;

$$k = (45.35 \times 10^8) \times \Phi_{ip}^{8.537,}$$

$$Sw = 0.02219 \times H^{-0.316} \times \Phi_{ip}^{-1.745}.$$

Class 2 comprises grain-dominated packstones, fine and medium crystalline grain-dominated dolopackstones, and medium crystalline mud-dominated dolostones;

$$k = (2.040 \times 10^6) \times \Phi_{ip}^{6.38} \text{ (recommended transform)},$$

$$Sw = 0.1404 \times H^{-0.407} \times \Phi_{ip}^{-1.440}.$$

Class 3 contains mud-dominated limestones and fine crystalline mud-dominated dolostones;

$$k = (2.884 \times 10^3) \times \Phi_{ip}^{4.275},$$

$$Sw = 0.6110 \times H^{-0.505} \times \Phi_{ip}^{-1.210}.$$

2.4.4
Petrophysics of Separate-Vug Pore Space

The addition of separate-vug porosity to interparticle porosity alters the petrophysical characteristics by altering the manner in which the pore space is connected, all pore space being connected in some fashion. Examples of separate-vug pore space are illustrated in Fig. 22. Separate vugs are not connected to each other. They are connected only through the interparticle pore space and, although the addition of separate vugs increases total porosity, it does not significantly increase permeability (Lucia 1983). Figure 23a illustrates this principle. Permeability of a moldic grainstone is less than would be expected if all the total porosity were interparticle and, at constant porosity, permeability increases with decreasing separate-vug porosity (Lucia and Conti 1987). The same is true for a large crystalline dolowackestone in that the data are plotted to the left of the class 1 field in proportion to the separate-vug porosity (Lucia 1983).

This principle is also true for intragrain microporosity. Fig. 23b shows data from a San Andres dolograinstone with intragrain microporosity from the Permian of West Texas. The plot shows that the permeability of the grainstone is less than would be expected if all the porosity were interparticle.

Separate vugs found within the oil column are usually considered to be saturated with oil because of their relatively large size. Oil migration into separate vugs is controlled by interparticle pore size and if the interparticle pore space is occluded with cement, the separate vugs may be water bearing. Intragrain microporosity, however, is unique because of small pore sizes. The small pore size produces high capillary forces that trap water and lead to anomalously high water saturations within a productive interval (Pittman 1971; Dixon and Marek 1990). Figure 24 compares the capillary characteristics of 2 Cretaceous grainstones composed of grains having microporosity between 2-µm crystals (Keith and Pittman 1983) and a San Andres dolograinstone having microporosity between 10-µm dolomite crystals with capillary characteristics of associated grainstones dominated by intergrain porosity. The capillary pressure curves are typically bimodal. This figure illustrates that within the oil column, water saturations in grainstones with intragrain microporosity are significantly higher than would be expected if no intragrain microporosity were present and all the porosity were intergrain.

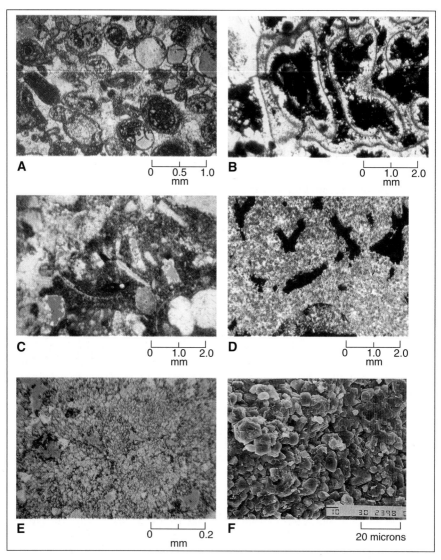

Fig. 22. Examples of vug pore types. Separate-vug types: **A** Oomoldic porosity, Ø = 26%, k = 3 md, Wolfcampian, West Texas. **B** Intrafossil pore space in a gastropod shell, Cretaceous, Gulf Coast. **C** Fossil molds in wackestone, Ø = 5%, k = 0.05 md. **D** Anhydrite molds in grain-dominated packstone, Ø = 10%, k = <0.1 md, Mississippian, Montana. **E** Fine crystalline dolograinstone with intergranular and intragranular microporosity pore types, Ø = 10%, k = 3 md, Farmer field, West Texas. **F** Scanning electron photomicrograph of dolograins in E showing intragranular microporosity between 10-µm crystals. Touching-vug types: **G** Cavernous porosity in a Niagaran reef, northern Michigan. **H** Collapse breccia, Ellenburger, West Texas. **I** Solution-enlarged fractures, Ellenburger, West Texas. (**J**) Cavernous porosity in Miami oolite, Florida. (**K**) Fenestral porosity in pisolitic dolostone. Note that the fenestral pores are more than twice the size of the enclosing grains

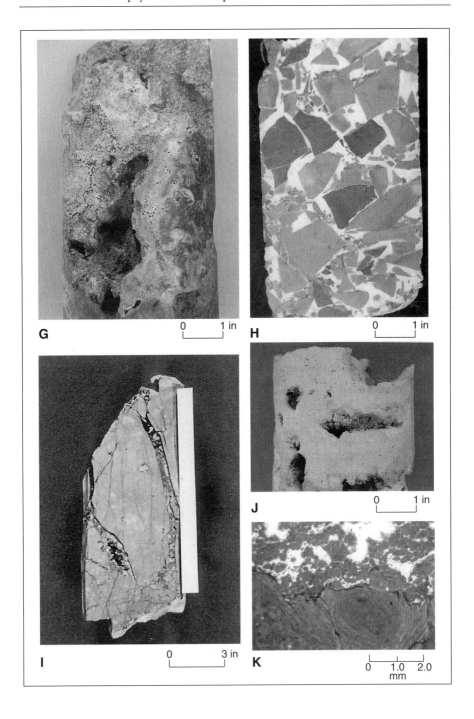

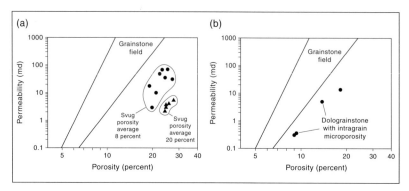

Fig. 23. Cross plot illustrating the effect of separate-vug porosity on air permeability. **A** Grainstones with separate-vug porosity (*Svug*) in the form of grain molds plot to the right of the grainstone field in proportion to the volume of separate-vug porosity. **B** Dolograinstones with separate vugs in the form of intragranular microporosity plot to the right of the grainstone field

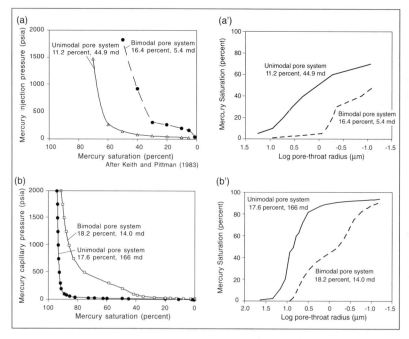

Fig. 24. Capillary pressure curves and pore-size distribution illustrating the effect of intragrain-microporosity on capillary properties. Note the bimodal character of the samples with both intergrain and intragrain microporosity pore types. **A** Ooid grainstone with intergrain porosity compared with ooid grainstone with both intergrain porosity and intragrain microporosity pore types (bimodal), Rodessa limestone, Cretaceous, East Texas (Keith and Pittman, 1983). **B** Dolograinstone with intergrain porosity compared with dolograinstone with both intergrain porosity and intragrain microporosity pore types (bimodal), West Texas

2.4.5
Petrophysics of Touching-Vug Pore Space

Examples of touching-vug pore types are illustrated in Fig. 22. Touching vugs can increase permeability well above what would be expected from the inter-particle pore system and are usually considered to be filled with oil in reservoirs. Lucia (1983) illustrated this fact by comparing a plot of fracture permeability versus fracture porosity to the three porosity/permeability fields for interparticle porosity (Fig. 25). This graph shows that permeability in touching-vug pore systems is related principally to fracture width and is sensitive to extremely small changes in fracture porosity. Unfortunately, there is no effective method for measuring fracture width or spacing using the rock fabric approach.

2.5
Summary

The goal of reservoir characterization is to describe the spatial distribution of petrophysical parameters such as porosity, permeability, and saturation. The rock fabric approach presented here is based on the premise that pore-size dis-

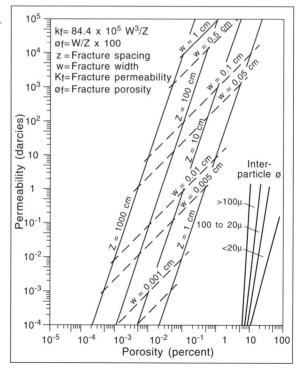

Fig. 25. Theoretical fracture air permeability-porosity relationship compared to the rock-fabric/petrophysical porosity, permeability fields (Lucia 1983). *W* Fracture width; *Z* fracture spacing

$k_f = 84.4 \times 10^5\ W^3/Z$
$\emptyset_f = W/Z \times 100$
z = Fracture spacing
w = Fracture width
K_f = Fracture permeability
$\emptyset_f$ = Fracture porosity

tribution controls the engineering parameters of permeability and saturation and that pore-size distribution is related to rock fabric, a product of geologic processes. Thus, rock fabric integrates geologic interpretation with engineering numerical measurements.

To determine the relationships between rock fabric and petrophysical parameters it is necessary to define and classify pore space as it exists today in terms of petrophysical properties. This is best accomplished by dividing pore space into pore space located between grains or crystals, called interparticle porosity, and all other pore space, called vuggy porosity. Vuggy pore space is further subdivided into two groups based on how the vugs are interconnected; (1) vugs that are interconnected only through the interparticle pore network are termed *separate vugs* and (2) vugs that are in direct contact with one another are termed *touching vugs*.

The petrophysical properties of interparticle porosity are related to particle size, sorting and interparticle porosity. Grain size and sorting of grains and micrite is based on Dunham's classification, modified to make it compatible with petrophysical considerations. Instead of dividing fabrics into grain support and mud support, fabrics are divided into grain-dominated and mud-dominated. The important attributes of grain-dominated fabrics are the presence of open or occluded intergrain porosity and a grain supported texture. The important attribute of mud-dominated fabrics is that the areas between the grains are filled with mud even if the grains appear to form a supporting framework.

Grainstone is clearly a grain-dominated fabric but Dunham's packstone class bridges an important petrophysical boundary. Some packstones, as we see them now, have intergrain pore space and some have the intergrain spaces filled with mud. Therefore, the packstone textural class must be divided into two rock-fabric classes, grain-dominated packstones that have intergrain pore space or cement and mud-dominated packstones where the intergrain spaces are filled with mud.

The important fabric elements to recognize for petrophysical classification of dolomites are precursor grain size and sorting, dolomite crystal size, and intercrystal porosity. Important dolomite crystal size boundaries are 20 and 100 μm. Dolomite crystal size has little effect on the petrophysical properties of grain-dominated dolostones. The petrophysical properties of mud-dominated dolostones, however, are significantly improved when the dolomite crystal size is >20 μm.

Permeability and saturation characteristics of interparticle porosity can be grouped into three rock-fabric/petrophysical classes. Class 1 is composed of (1) limestone and dolomitized grainstones and (2) large crystalline grain-dominated dolopackstones and mud-dominated dolostones. Class 2 is composed of (1) grain-dominated packstones, (2) fine to medium crystalline grain-dominated dolopackstones, and (3) medium crystalline mud-dominated dolostones. Class 3 is composed of (1) mud-dominated limestone and (2) fine crystalline mud-dominated dolostones.

Generic permeability transforms and water saturation, porosity, reservoir-height equations for each rock-fabric/petrophysical class are presented below.

Class 1;
$$k = (45.35*10^8) * \Phi_{ip}^{8.537},$$
$$Sw = 0.02219 * H^{-0.316} * \Phi_{ip}^{-1.745}.$$

Class 2;
$$k = (2.040*10^6) * \Phi_{ip}^{6.38} \text{ (recommended transform)},$$
$$Sw = 0.1404 * H^{-0.407} * \Phi_{ip}^{-1.440}.$$

Class 3;
$$k = (2.884*10^3) * \Phi_{ip}^{4.275},$$
$$Sw = 0.6110 * H^{-0.505} * \Phi_{ip}^{-1.210}.$$

The addition of separate-vug porosity to interparticle porosity increases total porosity but does not significantly increase permeability. Therefore, it is important to determine interparticle porosity by subtracting separate-vug porosity from total porosity and by using interparticle porosity to estimate permeability. Separate-vug porosity is normally considered to be filled with hydrocarbons in the reservoir. Intragrain microporosity, however, may contain significant amounts of capillary-bound water resulting in water free production of hydrocarbons from intervals with higher than expected water saturation.

Touching-vug pore systems cannot be related to porosity but are related principally to fracture width and continuity. Because there is no effective method for making this observation in the reservoir, the rock-fabric approach cannot be used to characterize touching-vug reservoirs.

The key to constructing a geologic model that can be quantified in petrophysical terms is to select facies or units that have unique petrophysical qualities for mapping. In non-touching vug reservoirs, the most important fabric elements to describe and map are (1) grain size and sorting using the modified Dunham classification, (2) dolomite crystal size using 20 and 100 μm as size boundaries, (3) interparticle porosity, (4) separate-vug type with special attention to intragrain microporosity, and (5) separate-vug porosity.

In touching vug reservoirs, characterizing the pore system is difficult because the pore system is not related to a precursor depositional fabric but is usually wholly diagenetic in nature. Whereas it may conform to bedding, as in the case of evaporite collapse brecciation and associated fractures, it more often cuts across strata. However, the recognition of the presence of a touching-vug pore system is paramount because it may dominate the flow characteristics of the reservoir.

Two basic steps in predicting the spatial distribution of petrophysical properties are (1) describing the one-dimensional distribution of rock fabrics and petrophysical properties from core and wireline log data and (2) extrapolating this information in three dimensions using geologic processes and stratigraphic principles. In the next chapter we will discuss describing rock fabrics and petrophysical properties in one dimension using core and wireline log data.

References

Alger RP, Luffel DL, Truman RB (1989) New unified method of integrating core capillary pressure data with well logs. Soc Pet Format Eval 4, 2: 145–152

Archie GE (1952) Classification of carbonate reservoir rocks and petrophysical considerations. AAPG Bull 36, 2: 278–298

Asquith GB (1986) Microporosity in the O'Hara oolite zone of the Mississippian Ste. Genevieve Limestone, Hopkins County, Kentucky, and its implications for formation evaluation. Carbonates Evaporites 1, 1: 7–12

Aufricht WR, Koepf EH (1957) The interpretation of capillary pressure data from carbonate reservoirs. Trans AIME 210: 402–405

Bebout DG, Lucia FJ, Hocott, CF, Fogg, GE, Vander Stoep GW (1987) Characterization of the Grayburg reservoir, University Lands Dune field, Crane County, Texas. The University of Texas at Austin, Bureau of Economic Geology, Report of Investigations No 168, 98 pp

Choquette PW, Pray LC (1970) Geologic nomenclature and classification of porosity in sedimentary carbonates. AAPG Bull 54, 2: 207–250

Choquette PW, Steiner RP (1985) Mississippian oolite and non-supratidal dolomite reservoirs in the Ste. Genevieve Formation, North Bridgeport Field, Illinois Basin. In: Roehl PO, Choquette PW (eds) Carbonate petroleum reservoirs. Springer, Berlin Heidelberg New York, pp 209–238

Dixon FR, Marek BF (1990) The effect of bimodal pore size distribution on electrical properties of some Middle Eastern limestones. Soc Petroleum Engineers Techn Conf, New Orleans, Louisiana, Sept 1990, SPE 20601

Dunham RJ (1962) Classification of carbonate rocks according to depositional texture. In: Ham WE (ed) Classifications of carbonate Rocks–a symposium. AAPG Mem 1: 108–121

Heseldin GM (1974) A method of averaging capillary pressure curves. Soc Professional Well Log Analysts Ann Logging Symp, June 2-5, paper E.

Keith BD, Pittman ED (1983) Bimodal porosity in oolitic reservoir–effect on productivity and log response, Rodessa Limestone (Lower Cretaceous), East Texas Basin. AAPG Bull 67, 9: 1391–1399

Kerans C (1989) Karst-controlled reservoir heterogeneity in the Ellenburger Group carbonates of West Texas. AAPG Bull 72, 10: 1160–1183

Kerans C, Lucia FJ, Senger RK (1994) Integrated characterization of carbonate ramp reservoirs using Permian San Andres Formation outcrop analogs. AAPG Bull 78, 2: 181–216

Leverett MC (1941) Capillary behavior in porous solids. Trans AIME 142: 151–169

Lucia FJ (1962) Diagenesis of a crinoidal sediment. J Sediment Petrol 32, 4: 848–865

Lucia FJ (1983) Petrophysical parameters estimated from visual description of carbonate rocks: a field classification of carbonate pore space. J Pet Technol March: 626–637

Lucia FJ (1993) Carbonate reservoir models: facies, diagenesis, and flow characterization. In: Morton-Thompson D, Woods AM (eds) Development geology reference manual. AAPG Methods in Exploration 10. AAPG, Tulsa, Oklahoma, pp 269–274

Lucia FJ (1995) Rock Fabric/petrophysical classification of carbonate pore space for reservoir characterization. AAPG Bull. 79, 9: 1275–1300

Lucia FJ, Conti RD (1987) Rock fabric, permeability, and log relationships in an upward-shoaling, vuggy carbonate sequence. The University of Texas at Austin, Bureau of Economic Geology, Geological Circular 87-5, 22 pp

Lucia FJ, Kerans C, Senger RK (1992a) Defining flow units in dolomitized carbonate-ramp reservoirs. Society of Petroleum Engineers Techn Conf, Washington D.C., SPE 24702, pp 399–406

Lucia FJ, Kerans C, Vander Stoep GW (1992b) Characterization of a karsted, high-energy, ramp-margin carbonate reservoir: Taylor-Link West San Andres Unit, Pecos County, Texas. The University of Texas at Austin, Bureau of Economic Geology, Report of Investigations No 208, 46 pp

Major RP, Vander Stoep GW, Holtz MH (1990) Delineation of unrecovered mobile oil in a mature dolomite reservoir: East Penwell San Andres Unit, University Lands, West Texas. The University of Texas at Austin, Bureau of Economic Geology, Report of Investigations No 194, 52 pp

Moshier SO, Handford CR, Scott RW, Boutell RD (1988) Giant gas accumulation in "chalky"-textured micritic limestones, Lower Cretaceous Shuaiba Formation, Eastern United Arab Emirates. In: Lomando AJ, Harris PM (eds) Giant oil and gas fields. Society of Economic Paleontologists and Mineralogists (SEPM) Core Workshop No 12, pp 229–272

Murray RC (1960) Origin of porosity in carbonate rocks. J Sediment Petrol 30, 1: 59–84

Pittman ED (1971) Microporosity in carbonate rocks. AAPG Bull 55, 10: 1873–1881

Pittman ED (1992) Relationship of porosity and permeability to various parameters derived from mercury injection-capillary pressure curves for sandstone. AAPG Bull 72, 2: 191–198

Scholle PA (1977) Chalk diagenesis and its relation to petroleum exploration: Oil from chalks, a modern miracle?. AAPG Bull 61, 7: 982–1009

Senger RK, Lucia FJ, Kerans C, Ferris MA (1993) Dominant control of reservoir-flow behavior in carbonate reservoirs as determined from outcrop studies. In: Linville B, Burchfield RE, Wesson TC (eds) Reservoir characterization III. Pennwell Books, Tulsa, Oklahoma, pp 107–150

Rock-Fabric/Petrophysical Properties from Core Description and Wireline Logs: The One-Dimensional Approach

3.1
Introduction

Petrophysical measurements and rock-fabric descriptions provide a basis for quantifying geological descriptions with petrophysical properties but represent point data. Point data are expanded in one dimension by detailed sampling of core material, and the vertical sequence of rock fabrics is used to extrapolate the petrophysical data laterally, as will be discussed in a later chapter. Core samples are normally available from only a few wells, whereas wireline logs are available from most wells. Therefore, most geological as well as petrophysical information must be derived from wireline logs.

A basic geological requirement for constructing a rock-fabric image of the reservoir is a 3-D stratigraphic framework. Construction begins with the one-dimensional description of stratigraphic relationships from core material and, most commonly, linking key stratigraphic surfaces to the gamma-ray log. A second basic requirement for quantifying the framework in petrophysical terms is knowledge of the spatial distribution of rock fabrics within the framework. This task also begins with the one-dimensional description of rock fabrics from core samples. Basic fabric parameters to describe are (1) interparticle porosity, (2) grain size and sorting, (3) dolomite crystal size, (4) separate-vug porosity, and (5) touching-vug pore space. These basic parameters are calibrated to wireline logs. Interparticle porosity is estimated by subtracting separate-vug porosity, as estimated from acoustic logs, from total porosity, as determined from neutron and density logs. Grain size and sorting may be estimated from the gamma-ray log, but resistivity and conductivity logs are better tools, especially when converted to water saturation. Dolomite crystal size can also be estimated from electric logs. Touching-vug pore space is best described from formation imaging logs, although acoustic logs are often useful.

The purpose of this chapter is to illustrate methods of obtaining rock-fabric data from wireline logs and integrating rock-fabric information into wireline-log calculations of porosity, water saturation, and permeability through a technique referred to as *core-log calibration*. Core description and analysis will be discussed first followed by methods of obtaining rock-fabric information from wireline logs.

3.2
Core Description

The first step in quantifying a geologic model is the description of rock fabrics from core material, using the Lucia classification methods described in Chapter 2. The descriptions are then calibrated with core-analysis data. The samples described should be the samples used to measure petrophysical properties. If other samples are used it will be difficult to match rock-fabric descriptions with core analysis data because of the high small-scale variability of porosity and permeability in carbonate reservoirs. The best sampling method is to drill 1-inch-diameter core plugs for analysis and use an end piece of the plug for rock fabric description. Thin sections are prepared from the end pieces for detailed description. If whole core samples are used in the analysis, a thin section is made from a sample of the whole core.

A key step in core-log calibration is depth shifting core depths to match log depth. Depths that are marked on core samples are notoriously inaccurate because the cores are often mishandled. It is common for sample depths shown on core analysis records to disagree with the depths scribed on the core. However, the core analysis number can often be found on core samples and should be used to identify the depth of the sample. If the core analysis sample number cannot be located, the depth scribed on the core will have to be used and the correlation between core description, core analysis, and core analysis will be only approximate.

The critical observations to record when describing the core for petrophysical quantification are lithology, grain size and sorting, dolomite crystal size, amount and type of separate vugs, and a description of touching vugs. Supplementary information includes grain types. This information can only be derived from visual observations of core slabs and thin sections. Additional information can be obtained from a high-resolution scanning electron microscope. A suggested spread sheet that includes the critical fabric information and the measured porosity and permeability is illustrated in Fig. 1.

Strata	Depth Sp. No.	LITHOLOGY							TEXTURE		PORE CLASS					CORE	
		Dolomite	Calc	Sulfate				Quartz	Acc.	Grain fraction	IPpor	Class	Size/Sort	Svug		Tvug	Por	Perm
		%	size	%	Anhy	type	Gyp	type	%	size	0 1	%		um	%	type	type	%	md

Fig. 1. Spread sheet for rock fabric/petrophysical descriptions

Rock fabric data is usually presented as a depth plot (Fig. 2) and the vertical successions of rock fabrics analyzed for predictable patterns. There is no standard method for constructing these rock-fabric depth plots. Methods that are generally accepted have been published by Bebout and Loucks (1984) and by Swanson (1981).

3.3
Core Analysis

Petrophysical data obtained from core measurements is often considered to be the truth against which log data should be compared. Routine core analysis data, however, can be very inaccurate for the reasons listed below.

1. Some methods of measuring porosity, such as the sum-of-fluids method, are very inaccurate. Only porosity values determined by using the Boyles Law method should be considered quantitative.
2. Biased sampling of the core can result in incorrect representations of porosity and permeability.
3. Permeability values obtained from routine whole-core analysis are typically too high on the low end and too low on the high end when compared with data from controlled laboratory procedures, probably because the samples are not properly sealed and insufficient care is taken with the measurements.
4. Induced fractures and stylolites commonly result in unrealistically high permeability values because insufficient confining pressure is used.
5. Porosity values can be too low due to incomplete removal of hydrocarbons or other fluids.
6. Porosity and permeability values can be too high if high-temperature procedures are used on core material containing gypsum and clay minerals, because high temperatures can significantly alter these minerals.

3.4
Core/Log Calibration

The second step in quantifying a geologic model is to correlate geologic descriptions with wireline log response. Key stratigraphic surfaces and rock fabrics obtained from core descriptions must be calibrated with wireline logs to expand the coverage of one-dimensional data. Wireline logs that have been used for this purpose include gamma-ray, neutron, density, acoustic, resistivity, dipmeter, and formation imaging logs. A general discussion of these logs and their function in quantifying a geologic model is presented below. Rock fabrics and basic rock-fabric attributes of porosity, interparticle porosity, separate-vug porosity, and particle size sorting can be calculated from these logs.

Wireline logs, however, respond to physical properties and not to the geologic attributes. A log cannot distinguish between grainstone and wackestone but can differentiate between levels of natural gamma-ray radiation, which may uniquely distinguish between grainstones and wackestones. A log does not measure differences in particle sizes, but it can measure rock resistivities, which relate to saturations and, in turn, to particle size and sorting. Wireline logs can distinguish different densities, which can be related to lithology. Such empirical relationships between log response and geological characteristics of the reservoir are used to extend the core data set throughout the field.

3.4.1
Procedures for Core-Log Calibration

The initial step in calibrating core data with wireline logs is to depth shift core depths to log depths. Two methods are used; (1) comparing the core gamma-ray log with the wireline gamma-ray log, and (2) comparing the porosity-depth plot with a porosity log, such as the neutron log. Accurate depth shifting of sample depths to core-analysis depths and core-analysis depths to wireline log depths is a tedious but very important task.

The second step is to input the geological data in numerical form and combine with core and log data for calibration analysis. The description form illustrated in Fig. 1 can be used for this task because most of the data can be recorded in numerical form. The integration of rock-fabric data into wireline log calculations result in (1) more accurate values of porosity, saturation, and permeability and (2) the ability to extract rock-fabric data from wireline logs to integrate petrophysical and geological data.

3.4.2
Gamma-Ray Logs and Depositional Textures

The gamma-ray log measures the natural radiation of uranium, potassium, and thorium. Potassium and thorium are concentrated in the minerals that constitute insoluble residue in carbonate rocks, such as rock fragments and clays. Uranium, however, is commonly concentrated in dolostone and is most likely related to diagenesis. Many carbonate facies can be correlated with current energy and thus to gamma radiation because the amount of insoluble residue is thought to be inversely proportional to current energy. Grainstones and grain-dominated packstones are deposited in high energy environments and typically have low gamma-ray activity. Mud-dominated packstones, wackestones and mudstones are deposited in low energy environments and typically have higher gamma-ray activity (Fig. 2). Other lithologies that have high gamma radiation are shaley intervals, quartz silt beds, and organic-rich beds. Sylvite (KCl) found associated with beds of halite (NaCl) has high gamma radiation from the potassium.

The uranium radiation must be removed from the gamma-ray signal before the gamma-ray log can be used effectively to identify rock fabrics because uranium is diagenetic in origin. The gamma-ray spectroscopy tool distinguishes between uranium, thorium, and potassium sources of radiation and provides a means of separating diagenetic uranium from depositional thorium and potassium (Fig. 3).

Fig. 2. Comparison of particle size and sorting from core description with gamma-ray log showing an 80% agreement (Grayburg/San Andres Reservoir, Farmer Field, West Texas). A 30 API gamma-ray value is used to distinguish grain-dominated from mud-dominated fabrics

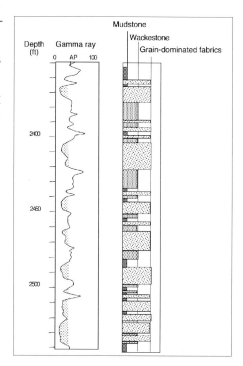

Fig. 3. Spectral gamma-ray log showing peaks caused by (1) high uranium content produced diagenetically and (2) high potassium and thorium content suggesting insoluble residue (siliciclastic) related to depositional energy conditions. Notice the response of the acoustic log to the presence of siliciclastic material. This is a lithology response that is often mistaken for high porosity

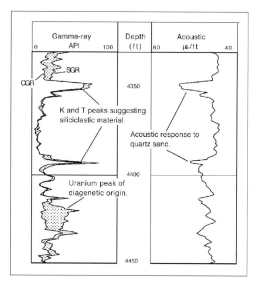

The gamma-ray log is displayed in API units, commonly from 0 to 150. Often it is necessary to reduce the scale of display to as small as 0 to 50 to observe subtle differences between the rock fabrics. The spectral gamma-ray log is displayed as two logs, the spectral gamma-ray log (SGR) combines all sources of radiation and the computed gamma-ray log (CGR) combines thorium and potassium sources. The CGR should be used for rock-fabric identification and lateral correlation of depositional events.

Repeat runs of gamma-ray logs in cased hole production wells have shown increased gamma-ray response over perforated intervals with time. This is probably the result of radioactive elements deposited along with well-bore scale that forms due to the flow of produced water through the perforated intervals.

3.4.3
Borehole Environment

Wireline log measurements respond to rock properties and the properties of the fluids in the pore space. The nature of the fluids in the pore space immediately surrounding the borehole depends upon the amount and type of mud filtrate that invades the formation. During the drilling process, drilling mud is pumped down the drill pipe and up the annulus between the drill pipe and formation. In formations with sufficient permeability, water will be filtered out of the mud and will infiltrate the formation, leaving mud cake on the side of the wellbore. This occurs because the pressure in the mud column is greater than the formation pressure. The invaded zone is composed of a flushed and a transition zone, and the dimensions of the transition zone will vary with time due to diffusion. In a water-bearing formation, the mud filtrate will completely flush the formation water near the well bore, and there will be a transition zone between the flushed zone and undisturbed formation (Fig. 4A). In a hydrocarbon-bearing formation composed of connate water and oil or gas, residual hydrocarbons will be trapped by capillary forces in the flushed zone, and there will be a transition zone between the flushed zone and the undisturbed reservoir (Fig. 4B).

The interpretation of wireline logs must consider the nature of the invaded zone. For neutron, density, and acoustic logs, the difference between oil and water is not large and is often ignored in calculations. However, the presence of residual gas in the invaded zone has a significant effect as will be discussed later. Water saturation and resistivity have a major effect on resistivity and conductivity logs. Micrologs and spherically focused logs are designed to measure the resistivity of the invaded zone (referred to as the Rxo), whereas deep-focused logs are designed to measure the resistivity of the uninvaded zone.

3.4.4
Neutron/Density Logs and Rock Fabric

Porosity is a basic value needed for volumetric calculations and for rock-fabric characterization. Porosity logs include neutron, density, and acoustic logs. The

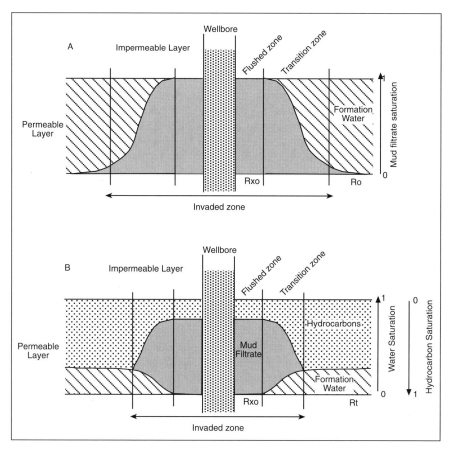

Fig. 4. Fluid composition of the invaded zone. In a permeable water-bearing formation, the mud filtrate will completely flush the formation water near the well bore, and there will be a transition zone between the flushed zone and the undisturbed formation (**A**). In a hydrocarbon-bearing formation composed of formation water and oil or gas, the flushed zone will be composed of mud filtrate and residual hydrocarbons trapped by capillary forces, the undisturbed reservoir will be composed of hydrocarbons and formation water, and the transition zone will be composed of hydrocarbons, mud filtrate and formation water (**B**)

neutron and density logs are designed to measure total porosity. The acoustic log is strongly affected by vuggy porosity and can be used to measure total porosity only in special situations, but can be used to quantify vuggy pore types. The neutron and density logs have their own unique inaccuracies and should be calibrated with core porosity values whenever possible. However, as described previously, core porosity values should not be accepted as "truth" until their accuracy is checked.

Total porosity alone can be calibrated to rock fabric in special cases. Compaction often reduces mud-dominated limestones to low porosity whereas compaction-resistant grain-dominated fabrics retain porosity and comprise the permeable reservoir rocks. In Paleozoic reservoirs, dolomite is commonly more porous than limestone, and dolomitization may selectively replace a specific fabric, such as wackestone (Lucia 1962). In these situations, porosity tools can be used to map a specific rock fabric.

Neutron logs measure the hydrogen ion concentration of the formation by measuring the capture of neutrons emitted by a neutron source. Common neutron tools are the compensated neutron and compensated dual spacing neutron logs, which are normally run in conjunction with a density log. This is a centralized tool that requires accurate corrections for borehole size; these corrections are usually automated into the final log display. The sidewall neutron log is a borehole pad device that requires less correction for borehole size and is more accurate. It is no longer available but is often encountered when working with old wells.

The neutron log responds to fluids in the invaded zone of the well bore because the depth of investigation is only a few inches. Water and residual oil and gas are present in the invaded zone and contain hydrogen ions. The hydrogen indexes of water and oil are similar and are usually assumed to be 1. Because of its much lower density, gas has a significantly lower concentration of hydrogen, resulting in a lower reading than expected from water or oil for the same porosity. If the gas saturation is not accounted for, the porosity values well be erroneously low. This is referred to as *the gas effect* and is commonly used to detect gas in combination with the density log described below.

Hydrogen ions are also present in some minerals as bound water. Common minerals that contain bound water are gypsum ($CaSO_4 \cdot 2H_2O$) and clay ($Al_2SO_5(OH)_4$). Organic material also contains hydrogen ions. The presence of these materials will result in erroneously high porosity values. Carbonate reservoir rocks do not usually contain enough clay or organic material to have a serious effect on the neutron log. Gypsum, however, can cause a serious problem for interpretation of neutron logs.

The following equation illustrates how common minerals found in carbonate reservoir rock effect the neutron porosity value, assuming no gas is present.

$$\phi n = 0.02V_d + 0.00V_c + 0.00V_a + 0.49V_g - 0.04V_q$$

(Pirson, 1983), where V_d, V_c, V_a, V_g, and V_q equal the bulk volumes of dolomite, calcite, anhydrite, gypsum, and quartz in the formation. This equation reveals that the mineral gypsum ($CaSO_4 \cdot 2H_2O$) will appear as 49% porosity on a neutron log. This large effect renders the neutron log ineffective as a porosity tool in the presence of gypsum (Fig. 5).

The *formation density log* emits medium-energy gamma rays, which lose energy when they collide with the electrons in the formation. The number of collisions is related to the bulk density of the formation, and the bulk density is related to porosity and common minerals found in carbonate reservoirs through the following equation.

$$\text{Bulk Density}(\rho_{bulk}) = \rho_f \phi + 2.71V_c + 2.84V_d + 2.98V_a + 2.35V_g + 2.65V_q$$

Fig.5. Depth plot overlaying CNL porosity and acoustic porosity logs showing the effect of bound water in gypsum on the CNL porosity (*shaded areas*)

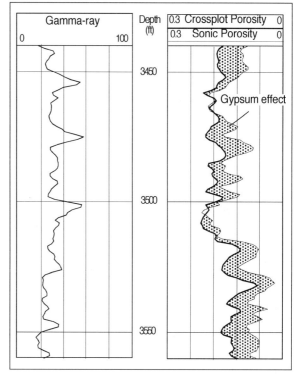

or

$$\phi = (\rho_{bulk} - \rho_{ma})/(\rho_f - \rho_{ma})$$

(Pirson, 1983), where ρ_f is the fluid density, ρ_{bulk} is the bulk density, ρ_{ma} is the matrix density, and V_c, V_d, V_a, V_g, and V_q equal the bulk volumes of dolomite, calcite, anhydrite, gypsum, and quartz in the formation.

The porosity can be calculated from the density log if the densities of the matrix and of the formation fluid are known. The standard method is to use a limestone density of 2.71 as matrix density and a fluid density of 1.1. A correction must be made if the formation is composed of minerals other than calcite, such as dolomite or anhydrite.

The density log is the basic lithology tool when used in conjunction with porosity values from the neutron log. The density log alone can be used to distinguish beds of anhydrite and halite from carbonate beds because they have unique densities of 2.97 and 2.0, respectively, and are typically dense (Fig. 6). In the presence of not more than two minerals, a cross plot of the density log and either the neutron or sonic log can be used to identify common minerals, such as dolomite and anhydrite or calcite and dolomite. Three minerals can be identified using all three porosity logs. *However, all mineral identification should be calibrated with local rock information to ensure accuracy.*

Fig. 6. Depth plot overlaying density and CNL porosity logs showing identification of halite, anhydrite, and dolomite beds. High gamma-ray values in dolomite beds suggest the presence of siliciclastic material

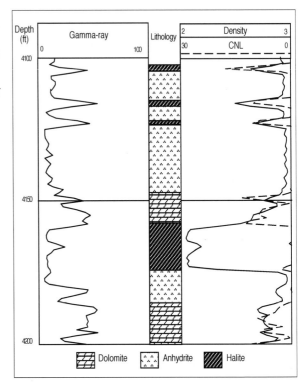

The formation fluid that affects the density tool is the fluid in the invaded zone of the well bore because of the small radius of investigation. Oil and gas have lower densities than water and their presence will result in lower bulk density values, and therefore erroneously high porosity values if fluid density is assume to be 1.1. The largest effect is due to the presence of residual gas in the invaded zone. This is referred to as *the gas effect* and is commonly used to detect gas in combination with the neutron log.

3.4.5
Acoustic Logs, Interparticle Porosity, and Vuggy Porosity

Acoustic logs record the travel time of compressional sound waves through an interval of the formation. The interval transit time is dependent upon lithology, porosity, and pore type. Large separate vugs tend to be overlooked by compressional acoustic waves. When large separate vugs are present, the interval transit time will be less than would be expected if all the porosity were interparticle, resulting in a porosity value lower than the true porosity. This does not appear to be true for small separate vugs such as intragrain microporosity.

The following equation, which is a form of the Wyllie time-average empirical formula, shows the relationship between acoustic log transit time (Δt), porosity (Φ), and acoustic transit time of common minerals (Δt_{ma}) found in carbonate reservoir rocks.

$$\Delta t = 189\phi_{tp} + (1-\phi_{ip})\Delta t_{ma}$$

or

$$\Delta t = \Delta t_f(\phi_{ip}) + (1-\phi_{tp})(44V_d + 49V_c + 50V_a + 52V_g + 56V_q)$$

(after Pirson, 1983), where Δt_f is fluid transit time, ϕ_{ip} is interparticle porosity, and $V_d, V_c, V_a, V_g,$ and V_q equal the bulk volumes of dolomite, calcite, anhydrite, gypsum, and quartz in the formation.

This equation is for nonvuggy carbonates and should not be used in the presence of more than a few percent of vuggy pore space. The equation shows that transit time has a large lithology effect and can be used to determine lithology in combination with other porosity logs. Siliciclastic beds can be easily identified in a carbonate sequence because of the large velocity difference between quartz and the carbonate minerals calcite and dolomite (Fig. 3). Care should be taken not to confuse high transit times with high porosity in mixed carbonate-clastic reservoirs. Dolomite reservoirs commonly contain sulfate in the form of anhydrite. However, the velocity difference between dolomite and anhydrite is not large enough to have a large effect on porosity calculations (Fig. 7). In shallow reservoirs, anhydrite is commonly converted to the hydrous form gypsum. As discussed previously, the hydrogen ions in gypsum render the neutron log virtually useless for calculating porosity. However, the small difference between the velocity of anhydrite and gypsum as expressed in the above equation does

Fig. 7. Cross plot of transit time and porosity for varying volumes of anhydrite, comparing the lithologic effect of anhydrite with the effect of separate-vug porosity. Variations in the volume of anhydrite in a dolomite cannot explain the behavior assigned to separate-vug porosity

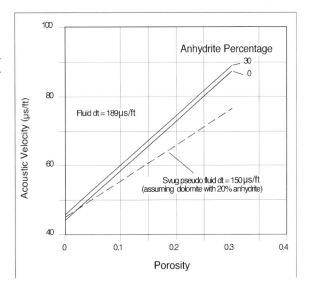

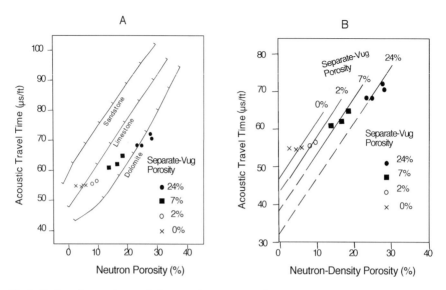

Fig. 8. Z-plot of acoustic travel time, total porosity, and percent separate-vug porosity from an oomoldic limestone. **A** Comparison with Schlumberger lithology lines illustrating how an oomoldic limestone could be mistaken for dolomite. **B** Parallel lines drawn through points with equal separate-vug porosity. The slope assumes a transit time equal to water (189 μs/ft)

not have a large effect on porosity calculations from acoustic logs. Therefore, acoustic logs can be used to calculate porosity in the presence of gypsum assuming the amount of separate-vug porosity is small (Bebout et al. 1987).

The effect of separate-vug porosity is illustrated by a Z-plot of acoustic transit time, total porosity calculated from porosity logs, and separate-vug porosity derived from thin section point counts of core samples. An example from an oomoldic limestone is shown in Fig. 8 (Lucia and Conti 1987) along with an example from an anhydritic dolomite (Fig. 9) (Lucia et al. 1995). In both cases the slope of the line relating transit time to porosity indicates a fluid transit time of about 150 μs/ft instead of the value for water of 189 ms/ft. Because no fluid has a transit time of 150 μs/ft, the lower transit time is assumed to be related to the presence of separate vugs.

Assuming a fluid velocity of 189 μs/ft, lines of equal separate-vug porosity can be defined and a transit time at zero porosity determined, referred to as a pseudo-matrix Δt (see Figs. 8 and 9). The pseudo-matrix Δt is defined in terms of Δt and porosity:

pseudo-matrix $\Delta t = \Delta t - 141.5 \times \phi$.

The pseudo-matrix Δt is plotted against the log of separate-vug porosity, and an equation relating separate-vug porosity, total porosity, and acoustic travel time is derived (Fig. 10). Three equations of this type are given below (Wang and Lucia 1993).

Fig. 9. Z-plot of acoustic travel time, total porosity, and percent separate-vug porosity from an anhydritic dolomite showing lines of equal separate-vug porosity assuming a fluid travel time of 189 μs/ft

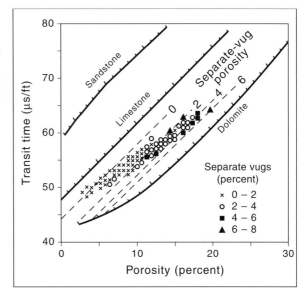

Fig. 10. Cross plot of the log of separate-vug porosity and pseudo-matrix transit times for oomoldic limestone and anhydritic dolomite illustrated in Figs. 8 and 9. The curves have similar slopes; separation is related to the difference in lithology

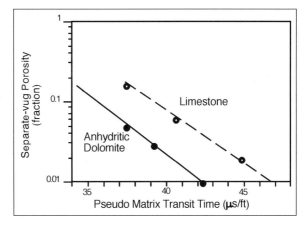

General,

$$\phi_{sv} = 10^{a-b(\Delta t - 141.5\phi)};$$

limestone,

$$\phi_{sv} = 10^{4.09-0.145(\Delta t - 141.5\phi)};$$

anhydritic dolomite,

$$\phi_{sv} = 10^{4.4419-0.15265(\Delta t - 141.5\phi)}.$$

These equations are suitable for large separate-vug pore types, such as moldic and intrafossil. They should not be applied to intragrain microporosity because experience has shown that this pore type is poorly manifested in acoustic, porosity cross plots.

3.4.6
Resistivity/Induction Logs and Rock Fabrics

Water saturations calculated from resistivity, induction, and porosity logs using the Archie equation can be used to estimate particle size and sorting. Water saturation is related to capillary pressure, which can be equated to reservoir height, and to pore-size distribution, which is a function of rock fabric. Water saturation is calculated using the Archie equation. Archie (1942) demonstrated that in 100% water-saturated rocks, the rock resistivity is related to (1) the amount of water present (the porosity), (2) the resistivity of the water, and (3) the pore geometry. These factors are related by the Archie equation given below.

Formation Resistivity Factor $F = R_O / \left(R_W \times \phi^{-m} \right)$,

where
R_o is the resistivity of 100% water-saturated rock,
R_w is the resistivity of formation water,
ϕ is the fractional porosity, and
m is the lithology exponent or cementation factor.

Archie (1942) also showed that, when hydrocarbons are present, the volume of water in the pore space is reduced and the resistivity is increased in proportion to the amount of hydrocarbons present. The resistivity of the formation in the presence of hydrocarbons is given by the following equation:

Resistivity Index $I = R_t / (R_o \times S_w^{-n})$,

Where
R_t is the resistivity of the rock containing the hydrocarbons,
R_o is the resistivity of 100% water saturated rock,
S_w is the fraction of pore space occupied by water, and
n is the saturation exponent.

Combining these two equations:

$Rt = R_w \phi^{-m} S_w^{-n}$

or

$S_w = (R_w / (R_t \times \phi^m))^{1/n}$.

Although saturation values are traditionally displayed as the fraction of pore space occupied by water, recently it has become convenient to display water saturation as *bulk volume water*, or the fraction of the rock occupied by water. The advantage of this display is that it more easily identifies intervals that will produce water. Bulk volume water (BVW) values higher than 0.04 are very likely to produce water (Asquith 1985).

Electric logs are used to measure the resistivity or conductivity of the formation. Three logs are typically run: a deep investigation, a shallow investigation, and a micro device that investigates the area immediately around the well bore. Borehole corrections are applied and a value for the true resistivity of the formation is calculated. Resistivity of the formation water is a function of salinity and temperature, and can be measured using produced waters and estimated from water saturation intervals using cross plots of porosity and resistivity extrapolated to 100% porosity (Pickett 1966).

The "m" value (lithology exponent or cementation factor) can be measured in the laboratory or estimated from wireline logs. It can be calculated from resistivity and porosity logs in water saturation intervals using the Archie equation if the water resistivity is known, or it can be estimated in hydrocarbon-bearing intervals from acoustic and porosity logs. Laboratory (Lucia 1983) and borehole (Lucia and Conti 1987) data has demonstrated that the "m" value is a function of the ratio of separate-vug porosity to total porosity, a ratio referred to as the *vug porosity ratio* (VPR; Fig. 11). This ratio can be calculated using separate-vug porosity estimated from acoustic logs, and total porosity calculated from neutron and density logs. A similar relationship has been developed by Brie et al. (1985) using a model composed of spherical pores.

The relationship between VPR and "m" values based on laboratory measurements and log calculations is shown in Fig. 11 and defined by the following equation.

Cementation Factor (m) = $2.14(\phi_{sv}/\phi_t) + 1.76$,

Fig. 11. Relationship between Archie m and the ratio of separate-vug porosity and total porosity (vug-porosity ratio, VPR) using laboratory data (Lucia, 1983) and log data. (Lucia and Conti, 1987)

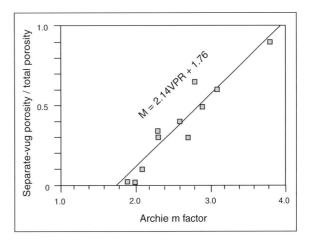

Where

ϕ_{sv} is the separate-vug porosity, and

ϕ_t is the total porosity.

The value of "m" for nontouching-vug carbonates ranges from 1.8 to as high as 4. In the presence of fractures and other touching-vug pore types, the "m" value may be less than 1.8 (Wang and Lucia 1993; Meyers 1991). Conventional wisdom suggests using an m value of 2 when no other information is available. However, if the range in m values is not properly accounted for in saturation calculations the resulting water saturations will be too low if m is larger than 2, and too high if m is smaller than 2. An example is shown in Table 1.

Table 1. The effect of "m" on water saturation calculations

R_t (ohm m)	Porosity	R_w	n	m	Calc S_w
400	0.2	1.6	2	2.0	32%
400	0.2	1.6	2	2.5	47%
400	0.2	1.6	2	3.0	71%

By changing the "m" value from 2 to 3, the water saturation changes from 32% to 71%, or from oil- to water-productive. Oomoldic grainstones, for example, have large m factors, and routine Archie calculations using an m of 2 typically result in low water saturations even though the rock is 100% water bearing.

Separate-vug porosity can be calculated from resistivity measurements if water saturation and total porosity are known, because resistivity is a function of the Archie m factor which is a function of separate-vug porosity. The pertinent equations for a water-bearing interval are given below (Lucia and Conti 1987):

$$R_O = R_W \phi_t^{-\left(2.14\left(\phi sv/\phi t\right)+1.76\right)} \text{ and}$$

$$\phi_{sv} = \left[\left(\log\left(R_W\right)-\log\left(R_O\right)\right)/\log\left(\phi_t\right)-1.76\right]\phi_t.$$

where
R_o is the resistivity of the water-saturated rock,
R_w is the resistivity of the formation water,
ϕ_{sv} is the separate-vug porosity, and
ϕ_t is the total porosity.

Water saturation is a function of pore-size distribution which is a function of particle size and sorting. The relationship between particle size and sorting, interparticle porosity, and water saturation based on capillary pressure curves is shown in Fig. 12A (Lucia 1995). An example of a similar plot based on wireline

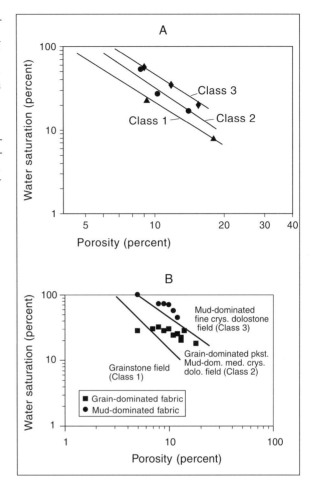

Fig. 12. Cross plots of porosity, water saturation, and rock fabric. **A** Cross plot of porosity and water saturation for three generic rock-fabric/petrophysical classes at a reservoir height of 150 m (500 ft) based on capillary pressure curves.
B Cross plot of porosity, water saturation, and rock fabrics from the Seminole San Andres field based on wireline log data. (Lucia et al. 1995)

log data from the Seminole San Andres Unit (Lucia et al. 1995) is shown in Fig. 12B. The effect of reservoir height is shown in Fig. 13 using generic rock-fabric specific saturation equations (see Chap. 2) and a typical porosity log from a Permian reservoir in West Texas. The three generic rock-fabric/petrophysical classes are easily distinguished up to 500 ft above the zero capillary pressure level, and the distinction becomes more apparent at lower reservoir heights. A saturation curve for a typical upward-shoaling cycle illustrates how particle size and sorting can be identified from porosity, water saturation, and reservoir height (Fig. 13B).

3.4.7
Formation Imaging Logs and Vuggy Pore Space

Formation imaging tools are an important source of information on touching-vug pore types. Modern tools are resistivity based, although the original tools

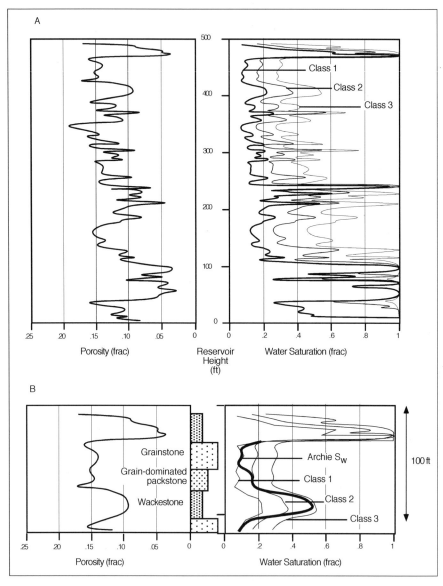

Fig. 13. Illustration of the effect of reservoir height and petrophysical class on water saturation. **A** A depth plot of water saturation for various classes based on the generic equations and porosity curve. Note that the three generic classes can be distinguished up to 500 ft. **B** A portion of A showing how the Archie saturations can be used to identify a vertical sequence of rock fabrics

were based on acoustic measurements. Schlumberger developed the first tool (the FMS) that has evolved into the FMI tool. The FMI tool has four orthogonal arms with two pads, each giving 60 to 100% coverage of the borehole. Each pad

Fig. 14. Example of an image log of collapse breccia. Microresistivity measurements from an array of 24 resistivity buttons on two pads of the tool are shown on the left. The microresistivity measurements are converted into the images of collapse breccia shown on the right. The light colored areas are the breccia blocks and the dark colored areas are a siliciclastic breccia matrix. (Hammes 1997)

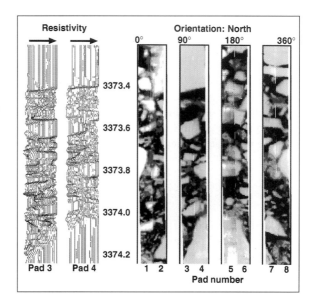

contains 24 microresistivity buttons resulting in a resolution of 0.3 in. The tool provides borehole drift and direction as well as formation images. Since the introduction of the FMI tool, other companies have developed similar logs.

The image logs can provide images of fractures, breccias, large vugs, and sedimentary structures useful for correlation and characterizing touching-vug pore systems (Fig. 14; Hammes 1997). The images, however, should be calibrated with core descriptions to ensure accuracy. The fluid in the invaded zone, formation sloughing, enlargement of the borehole, and tool sticking can affect the images.

The well survey data can be used to calculate dip and strike of planar fractures and bedding surfaces. Dips and strikes from the fore slope beds of carbonate buildups have been used to assist in locating pinnacle reefs.

3.4.8
Permeability from Wireline Logs, the Rock Fabric Method

Permeability cannot be obtained directly from wireline logs. The routine method is to correlate core permeability and porosity measurements and to use the resulting porosity-permeability transform to calculate permeability from porosity logs. This method too often averages out the robust permeability variations that are characteristic of carbonate reservoirs. Although there are statistical techniques for recreating the variability, these techniques do not spatially distribute the permeability values.

A more accurate method is to use interparticle porosity and rock-fabric-specific porosity-permeability transforms. Interparticle porosity is estimated by subtracting total porosity as calculated from neutron and density logs from separate-vug porosity as estimated from acoustic logs. Petrophysically significant rock-fabric classes can be determined from cross plots of water saturation, porosity, and reservoir height. More accurate saturation calculations can be obtained by integrating a variable m value based on a VPR calculated from the acoustic log. In reservoirs with a history of water flooding, the saturation calculations should be compared with estimates of original saturations using rock-fabric-specific saturation, porosity, and reservoir-height relationships. In some reservoirs, the gamma-ray log can be closely correlated with rock fabrics and used to improve permeability estimates.

3.5
Summary

The first step in constructing a reservoir model is to describe important stratigraphic and rock-fabric attributes seen in core material and relating the rock fabrics to petrophysical data measured on the core samples. Because few cores are taken and most wells have wireline logs, the stratigraphic and rock fabric information obtained from core description must be calibrated to wireline logs to increase the distribution of data over the field. The procedure starts with core description, moves to petrophysical calibration of descriptions, and finally to core-log calibration.

The logs commonly used to obtain geologic information include gamma-ray, neutron, density, acoustic, resistivity, dipmeter, and formation imaging logs. The gamma-ray log is the most common log used by geologists to correlate significant stratigraphic surfaces, but has limited use in identifying rock-fabric facies. The neutron log is the principal porosity tool and is used in conjunction with other tools to extract geological information. The density log is the principal tool for extracting mineralogical information.

Acoustic, resistivity and formation imaging logs are the best tools for gathering rock fabric information. The acoustic log is sensitive to lithology and to vuggy porosity. Permeability is related to interparticle porosity and it is necessary to have an estimate of separate-vug porosity to deduct from total porosity in order to estimate interparticle porosity. In the presence of large separate vugs, such as grain molds, separate-vug porosity can be estimated using an equation with the following form:

$$\phi_{sv} = 10^{a-b\left(\Delta t - 141.5\phi\right)}$$

Resistivity and induction logs are used to calculate water saturation using the Archie equation. Accurate values of the cementation factor m are required to calculate water saturations from resistivity logs. The m value is a function of the vug-porosity ratio which can be estimated using separate-vug porosity calculated from acoustic logs, and total porosity using the following equation:

cementation factor (m) = $2.14(\phi_{sv}/\phi_t) + 1.76$.

Water saturation is a function of reservoir height and pore-size distribution, and pore-size is a function of particle size and sorting. Cross plots of water saturation and porosity can be used to identify rock-fabric/petrophysical classes.

Touching-vug pore systems are important in reservoirs because they may have small volumes but can dominate fluid flow characteristics. Identification of touching-vug pore systems is best accomplished using formation-imaging logs.

Improved permeability estimates are made by using rock-fabric-specific transforms and interparticle porosity. The method is as follows.
1. Calculate total porosity using available porosity logs.
2. Calculate separate-vug porosity using travel-time/porosity/separate-vug relationships.
3. Calculate interparticle porosity by subtracting separate-vug porosity from total porosity.
4. Determine the rock-fabric/petrophysical class using either the gamma-ray log or rock-fabric specific water-saturation/porosity/height relationships compared with log calculated water saturation and porosity.
5. Calculate permeability by using rock-fabric-specific permeability transforms and interparticle porosity.

Improved water saturation estimates can be made by using the VPR to estimate the Archie m value. The method is as follows:
1. Calculate total porosity using available porosity logs.
2. Calculate separate-vug porosity using travel-time/porosity/svug relationships.
3. Calculate the vug-porosity ratio and use the generic equation to estimate the Archie m value.
4. Calculate the Archie water saturation using the Archie m value, estimate of n value, formation resistivity, and water resistivity.

Table 2 summarizes the rock fabric information that can be obtained from wireline logs.

Table 2. Calculation of rock-fabric petrophysical parameters from wireline logs

Input	Output
Total porosity	Separate-vug porosity
Transit time	Interparticle porosity
True resistivity	Petrophysical class
R_w	Archie m value
Lithology	S_w from variable m values
Saturation exponent (n)	Rock-fabric permeability

References

Alger RP, Luffel DL, Truman, RB (1989) New unified method of integrating core capillary pressure data with well logs. SPE Format Eval 4, 2: 145–152

Archie GE (1942) The electrical resistivity log as an aid in determining some reservoir characteristics: Trans AIME 146: 54–62

Asquith GB (1985) Handbook of log evaluation techniques for carbonate reservoirs. Methods in Exploration Series No. 5. AAPG, Tulsa, Oklahoma, 47 pp

Bebout DG, Loucks RG (1984) Handbook for logging carbonate rocks. The University of Texas at Austin, Bureau of Economic Geology, Handbook 5, 43 pp

Bebout DG, Lucia FJ, Hocott CF, Fogg GE, Vander Stoep GW (1987) Characterization of the Grayburg reservoir, University Lands Dune field, Crane County, Texas. The University of Texas at Austin, Bureau of Economic, Geology Report of Investigations No 168, 98 pp

Brie A, Johnson DL, Nurmi RD (1985) Effect of spherical pores on sonic and resistivity measurements. Society of Professional Well Loggers Association 26th Annu Logging Symp, Paper W, Houston, Texas, June 17–20, 20 pp

Dewan JT (1983) Essential of modern open-hole log interpretation. PennWell Books, Tulsa, Oklahoma, 361 pp

Hammes U (1997) Electrical imaging catalog: microresistivity images and core photos from fractured, karsted, and brecciated carbonate rocks. The University of Texas at Austin, Bureau of Economic Geology, Geological Circular 97-2, 20 pp

Hurd BG, Fitch JL (1959) The effect of gypsum on core analysis results. J Petrol Technol 216: 221–224

Lucia FJ (1962) Diagenesis of a crinoidal sediment. J Sediment Petrol 32, 4: 848–865

Lucia FJ (1983) Petrophysical parameters estimated from visual descriptions of carbonate rocks: a field classification of carbonate pore space. J Pet Technol, March 1983: 629–637

Lucia FJ (1995) Rock-fabric/petrophysical classification of carbonate pore space for reservoir characterization: AAPG Bull 79, 9: 1275–1300

Lucia FJ, Conti RD (1987) Rock fabric, permeability, and log relationships in an upward-shoaling, vuggy carbonate sequence. The University of Texas at Austin, Bureau of Economic Geology, Geological Circular 87-5, 22 pp

Lucia FJ, Kerans C, Wang FP (1995) Fluid-flow characterization of dolomitized carbonate ramp reservoirs: San Andres Formation (Permian) of Seminole field, and Algerita Escarpment, Permian Basin, Texas and New Mexico. In: Stoudt EL, Harris PM (eds) Hydrocarbon reservoir characterization: Geologic framework and flow unit modeling. SEPM Short Course no. 34, pp 129–155

Meyers MT (1991) Pore combination modeling: a technique for modeling the permeability and resistivity properties of complex pore systems. Soc Petroleum Engineers Techn Conf, Dallas, Texas, SPE 22662, p 77–88

Pickett GR (1966) A review of current techniques for determination of water saturation from logs. J Pet Technol November: 1425–1433

Pirson JS (1983) Geologic well log analysis. Gulf Publishing Company, Houston, Texas, 475 pp

Swanson RG (1981) Sample examination manual. AAPG Methods in Exploration Series. AAPG, Tulsa, Oklahoma, 65 pp

Wang FP, Lucia FJ (1993) Comparison of empirical models for calculating the vuggy porosity and cementation exponent of carbonates from log responses. The University of Texas, Bureau of Economic Geology, Geological Circular 93-4, 27 pp

Origin and Distribution of Depositional Textures and Petrophysical Properties: The Three-Dimensional Approach

4.1
Introduction

Petrophysical properties can best be distributed in the interwell space if constrained by a chronostratigraphic framework.

Petrophysical data obtained from wireline logs and cores are one-dimensional and have no intrinsic spatial connotation other than the frequency and spacing of the data. To construct a three-dimensional image, the interwell volume must be filled with data using a correlation method. Because the volume of the reservoir traversed by wellbores is about one millionth of the total rock volume, 99.99% of the reservoir model is dependent upon the method used to correlate petrophysical data between wells. Geostatistical methods are available for filling the space between wells with a heterogeneous pattern of petrophysical properties, and this method will be discussed in a later chapter. Whereas geostatistical methods are an improvement over using a few average properties to construct a model, the resulting images are unrealistic and produce unreliable performance predictions. The requirements for constructing a realistic model are (1) a chronostratigraphic geological framework and (2) a knowledge of geologic processes that form and modify petrophysical properties.

Key steps in constructing a petrophysical model are (1) relating rock fabrics to petrophysical parameters, (2) identifying the geological processes that formed the rock fabrics, (3) describing a cycle-based sequence stratigraphic framework, and (4) distributing petrophysically significant rock-fabric bodies within the stratigraphic framework. In Chapter 2 we discussed the relationship of rock fabrics to pore-size distribution and to petrophysical parameters and identified key fabric elements to describe. In Chapter 3 we discussed the description of 1-D vertical successions of rock fabrics from cores and the relationship between rock fabrics and petrophysical properties to wireline log measurements. In this chapter we will discuss the origin of depositional textures, the petrophysical properties of carbonate sediments, and the sequence stratigraphic framework. In the following chapters we will discuss the formation of carbonate reservoirs through diagenetic overprinting of the depositional texture and the resulting distribution of rock fabrics and petrophysical properties.

4.2
Textures, Mineralogy, and Petrophysical Properties of Carbonate Sediments

The initial process in the formation of a carbonate reservoir is sedimentation. Carbonate sediments are commonly produced in shallow, warm oceans either by direct precipitation out of seawater or by biological extraction of calcium carbonate from sea water to form skeletal material. The result is sediment composed of particles with a wide range of sizes, shapes, and mineralogies mixed together to form a multitude of textures, chemical compositions and, most importantly, associated pore-size distribution.

   Carbonate sediments can be divided into loose sediment and sediment bound together as a result of organic activity. Sediment can be organically bound by filamentous algae to form algal stromatolites, or by encrusting organisms such as the modern coralline alga *Lithothamnium* or Devonian stromatoporoids to form reefs. The binding action can create large constructional cavities, which result in highly permeable sediment (Fig. 1).

   More commonly, carbonate sediments are composed of loose grains. The grain size of the loose sediment is generally bimodal, being composed of a sand-sized (or larger) fraction and a mud-sized fraction. In general, sand-sized carbonate grains reflect the size of the calcareous skeletons or their calcified hard parts. Coral fragments, for example, are typically boulder-sized particles, whereas

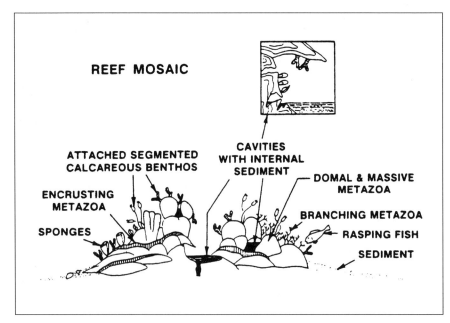

Fig. 1. Diagram of a reef assemblage showing large constructional cavities. The large constructional cavities and large coral fragments produce a depositional texture with large pore sizes. (after James 1984)

shell material is typically much smaller. Skeletal sediment undergoes various degrees of mechanical, biological, and chemical breakdown into smaller particles caused by exposure to strong marine currents, boring by fungi and algae, browsing by animals searching for food, and disintegration of organic material (Fig. 2). Mud-sized particles (Fig. 3) are formed principally by organisms, such as small calcareous plankton and mud-sized aragonite crystals produced by calcareous algae. Aragonite mud is also produced by precipitation from sea water, and white patches of aragonite mud, referred to as "whitings" are often seen floating in the ocean where carbonate sediment is being deposited (Milliman et al. 1993).

Biological and chemical processes can increase as well as decrease the particle size. Burrowing organisms pass mud-sized sediment through their digestive tract and produce sand-sized fecal pellets (Fig. 4), thus creating sand-sized particles out of mud-sized particles. Grain size can be enlarged by algal coatings

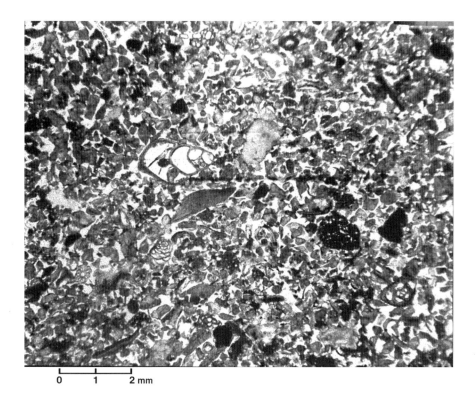

Fig. 2. Photomicrograph of a modern skeletal grainstone composed of coral, mollusks, coralline algae, and foram fragments. The size of the intergranular pores is controlled by the size of the grains. Note intragrain pore space within forams and gastropod

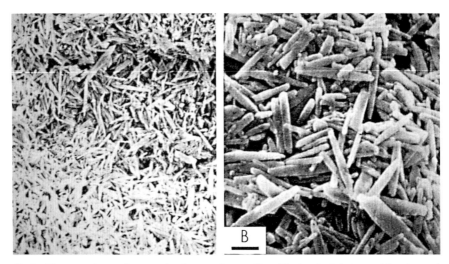

Fig. 3. Electron photomicrograph of aragonite needles. The bar in **B** is 1 μm. The size and shape of the pore space between the needles are controlled by the size and shape of the needles, and the pore sizes are very small because the particles are very small. (after Gebelein et al. 1980)

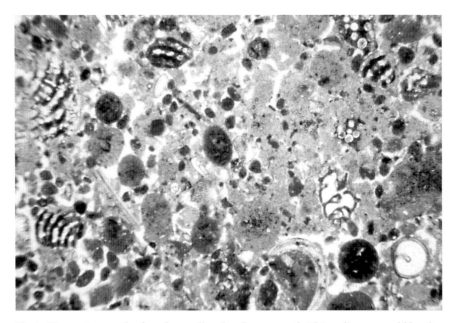

Fig. 4. Photomicrograph of modern pelleted carbonate mud. This sediment would be classified as a grain-dominated packstone because of the intergrain pore space. However, compaction may destroy the intergrain pore space, and the resulting fabric would be classified as a mud-dominated packstone or peloid wackestone. Photograph is 4.5 mm wide (courtesy of R.B. Perkins)

forming oncoids. Forams can bind mud and pellets together to form grapestone particles. Carbonate particles that are related to chemical processes include ooids, which are formed by chemical precipitation around a nucleus grain in the presence of strong current action; pisolites, which are a product of subaerial exposure (Gerhard 1985); and intraclasts, which are produced early by the breakup of lithified sediment in either the subaerial or subtidal environment.

Dunham's (1962) classification (Fig. 5) divides carbonates into organically bound sediments and loose sediments. A key consideration in Dunham's classification of loose sediment is the bimodal texture of carbonate sediment. Sediments are classified first by mud vs. grain support and then by amount of mud in the sediment. Generic names, such as wackestone and grainstone, are modified with grain type such as "trilobite wackestone" or "ooid grainstone."

Dunham's boundstone class was further divided by Embry and Klovan (1971) because carbonate reefs are commonly composed of large reef-building organisms, such as corals, sponges, and rudists, that are either bound together or transported, forming sediments composed of very large particles. They devised a method of describing the complex textural relationships, introducing the terms bafflestone, bindstone, and framestone to describe autochthonous (in-place) boundstone reef material, and the terms floatstone and rudstone to describe allochthonous transported reef sediment with particles larger than 2 mm in diameter. Rudstone is grain-supported whereas floatstone is mud-supported sediment.

This method of describing carbonate textures can also be used to describe pore geometries in carbonate sediments because "grain-supported" implies

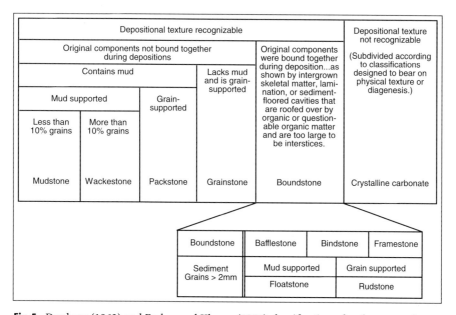

Fig. 5. Dunham (1962) and Embry and Klovan (1971) classification of carbonate rocks

intergrain pore space between sand-sized grains whereas "mud-supported" implies microporosity between mud-sized particles. However, petrophysical descriptions should divide the packstone class into mud-dominated and grain-dominated packstones, as described in Chapter 2. The pore-size distribution in bafflestones and bindstone is often not related to particle size and shape, and the pore types may best be grouped in the touching-vug class. The pore-size distribution of floatstone and rudstone depends upon the texture of the intergrain volume. If the intergrain volume is composed of lime mud (floatstone), the size of the mud particles will control pore size. If the intergrain volume is partially filled with or devoid of mud (rudstone), the size of the grains and intergrain mud will control the pore size.

The amount of porosity and the pore-size distribution of Holocene sediment is related to the size, shape, and internal structure of the sediment particles. Carbonate particles have various shapes, such as spherical ooids, flat and coiled shell fragments, platy and rod-shaped algae, and needle-shaped aragonite crystals, and various sizes such as 5-mm aragonite crystals, sand-sized ooids, and boulder-sized coral fragments. Pore space is found not only between carbonate grains and crystals but also within grains, such as in the living chambers of shells. Pore space located within shells and other grains is connected only through the interparticle pore space and is a type of separate-vug porosity. These factors result in a large variability of porosity and pore-size distribution in unconsolidated carbonate sediments. *However, the intergrain pore size and pore-size distribution is always a function of the size, shape, and sorting of particles.*

Enos and Sawatsky (1981) measured the porosity and permeability of modern carbonate sediments. Their results show a general increase in porosity as the mud fraction increases (Fig. 6a). Average porosity of grainstone is about

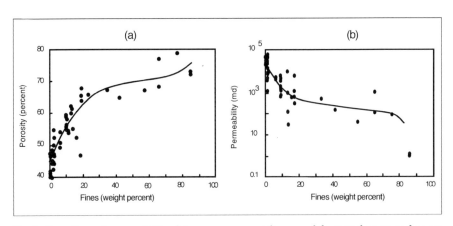

Fig. 6. Porosity and permeability data versus percent lime mud from Holocene carbonate sediments of the Bahamas and the Florida Keys. **a** Porosity increases as the percent fines increase to a value of 70%. **b** Permeability decreases as the percent fines increase because of the small pore size associated with mud-sized particles. (Enos and Sawatsky 1981)

45%, similar to extremely well sorted sandstone. Porosity increases up to 70% with increasing amounts of carbonate mud, a change that can be related to the open stacking of needle-shaped aragonite crystals found in carbonate mud (Fig. 3).

Permeability is a function of pore-size distribution, which is directly related to the porosity, particle size, and sorting in Holocene sediments. The permeability of mud-dominated sediments is between 1 and 200 md because of the small size of the pores between the 5-10 µm aragonite crystals (Fig. 6b). Permeability increases as the volume of mud decreases from about 20 to 0%, indicating an increasing influence of grain size on pore size. The permeability of grain-dominated sediment with some intergrain mud averages about 2000 md whereas grainstone averages about 30000 md.

A comparison of the porosity and permeability between Holocene sediments and carbonate reservoirs reveals that modern sediments have much higher porosity and permeability. Average porosity for reservoirs in the United States is 12% and average permeability is about 50 md (Schmoker et al. 1985) whereas carbonate sediments have porosity values higher than 40% and typical permeability values higher than 100 md. Therefore, all modern carbonate sediments have reservoir quality permeability.

An important consideration in diagenesis is the mineralogical composition of the carbonate sediment. Carbonate sediments are composed of three forms of carbonate that are variably stable (Walter 1985; Fig. 7). Aragonite has an orthorhombic crystal structure and is unstable at surface conditions. Calcite has a triclinic crystal structure and is stable at surface conditions in its pure form. However, magnesium can substitute for calcium in the crystal lattice, and stability decreases with increasing amounts of magnesium. Holocene calcite has as much as 20% magnesium carbonate; calcite with significant magnesium substitution is referred to as high-magnesium calcite.

The mineralogy of marine organisms varies with phyla and over geologic time (Fig. 8). For instance, modern red algae are composed of high-magnesium

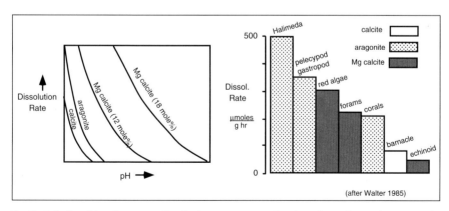

Fig. 7. Relationships between dissolution, texture, and mineralogy for carbonate sediments. (after Walter 1985)

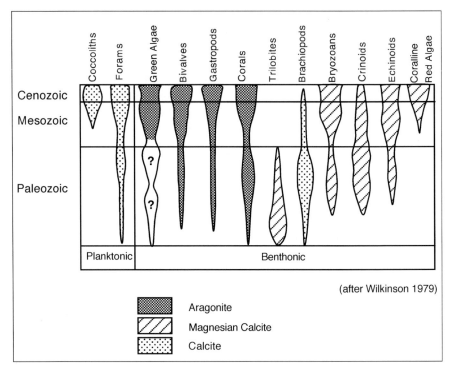

Fig. 8. Mineralogy of marine organisms. (Wilkinson 1979)

calcite, brachiopods are composed of low-magnesium calcite, and coral is composed of aragonite. Most modern ooids are composed of aragonite, but many ancient oolites were composed of low-magnesium calcite. Carbonate mud is normally composed of a mixture of the three minerals, aragonite commonly being the most abundant.

4.3
Spatial Distribution of Textures and Petrophysical Properties

The spatial distribution of petrophysical properties in carbonate sediments is controlled by the distribution of depositional textures, which is best described in terms of a chronostratigraphic, sequence stratigraphic framework composed of high-frequency cycles (HFC), high-frequency sequences (HFS), and composite sequences (Kerans and Fitchen 1995). An HFC is a chronostratigraphic unit defined as a succession of genetically related textures (beds or bedsets) bounded by marine flooding surfaces and their correlative surfaces. An HFS is an unconformity-bound series of HFCs that show transgression and progradation. A composite sequence is an unconformity-bound depositional sequence composed of multiple HFSs.

The high-frequency cycle concept has direct application to reservoir characterization and flow-modeling studies in carbonate reservoirs and is the most significant stratigraphic element in the stratigraphic framework. The upward and lateral succession of rock fabrics within the HFC can be translated into predictable patterns of petrophysical values that aid in log interpretation and in quantifying the framework in petrophysical terms for performance prediction.

4.3.1
High-Frequency Cycles and Facies Progression

High-frequency cycles are defined in one dimension by vertical successions of depositional textures and in two dimensions by lateral facies progressions. The two classic vertical successions that define cycles are (1) a subtidal cycle composed of a succession of textures that increase in grain size and sorting upward, and (2) a tidal-flat-capped succession of subtidal textures.

The vertical succession of texture reflects changes in current energy, and cycle thickness reflects accommodation space and sedimentation rate. The lateral distribution of depositional facies reflects energy levels, topography, and organic activity. These changes can be related to the geometry of the carbonate platform. Carbonate platform geometry has been divided into ramps and rimmed shelfs (Reed, 1985). Ramps have a gently dipping profile (0 to about a 2° dip; Fig. 9) and depositional rimmed shelves are those that have actively aggraded to form a steep shelf margin (15° dip to near vertical). Ocean currents are produced by tides and waves and are concentrated at major topographic features, such as ramp and rimmed shelf margins, islands, and shorelines (Fig. 9B). The combination of topography and ocean currents produces a facies progression from landward to basinward of (1) peritidal mud- to grain-dominated textures and evaporite facies, (2) middle-shelf mud-dominated and occasional grain-dominated packstone facies, (3) shelf-crest grain-dominated facies and reefs, (4) outer-shelf mud- to grain-dominated facies, and 5) basinal mud-dominated facies and debris flows (Figs. 9a and 10).

The peritidal facies is composed of tidal-flat capped cycles and is normally found in the most landward position of a carbonate shelf. The cycles are formed by filling accommodation space to sea level and deposition of sediment above sea level by transport of carbonate sediment onto the mud flat by tidal and storm currents. Therefore, tidal-flat capped cycles demonstrate a seaward movement of shoreline by the progradation of intertidal and supratidal sediments over subtidal sediments (Fig. 11B).

Tidal-flat capped cycles are normally formed where the shoreline is sheltered from wave action, and the subtidal sediments are dominated by carbonate mud with grains concentrated in channels and beach ridges. The tidal flat environment is divided into the intertidal zone, which is defined as the vertical interval between mean-high and mean-low tides and is covered by seawater twice a day, and the supratidal zone, which is defined as the area beyond the reach of the daily tides and is covered by normal sea water only during spring and storm tides.

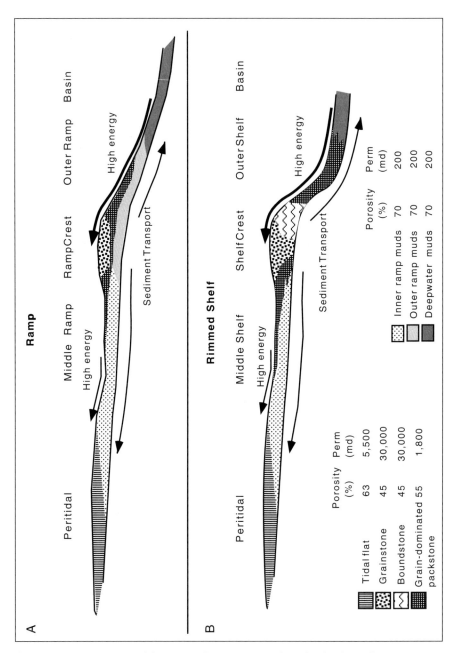

Fig. 9. Facies progression of depositional environments from land to basin for **A** ramp profile and **B** rimmed-shelf profile. The facies are systematically arranged based on topography and current energy. Characteristic Holocene porosity and permeability values for depositional textures from Enos and Sawatsky (1981) are listed

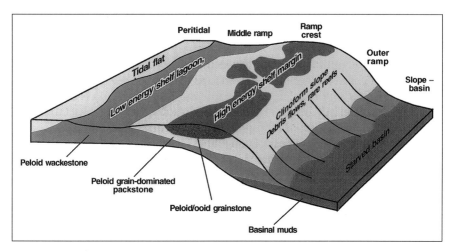

Fig. 10. Generalized block diagram illustrating carbonate-ramp facies patterns and topography

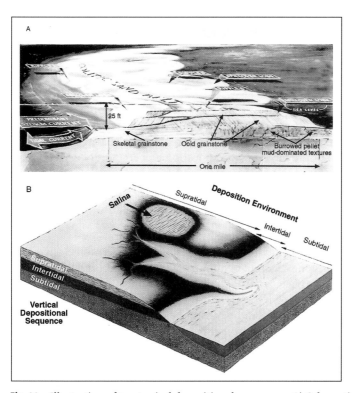

Fig. 11. Illustration of two typical depositional sequences. **A)** Schematic block diagram of the Cat Cay oolitic sand belt, Bahama Bank showing a vertical sequence from burrowed pelleted mud to ooid grainstone (Ball 1967). **B** Diagrammatic view of a prograding evaporitic tidal flat showing general depositional environments and vertical sequence. (Lucia 1972)

Sediment of the intertidal zone is characterized by burrowed, pelleted muddy sediment with no characteristic sedimentary structures and is best identified by its location immediately below a supratidal unit (see Table 1). Algal laminates are concentrated at the boundary between the intertidal and supratidal zones. The supratidal zone, sometimes referred to as the Sabkha referring to the extensive Persian Gulf supratidal flats (Patterson and Kinsman 1981) is easily recognized by characteristic irregular laminations, pisolites, mud cracks, intraclasts, and fenestral fabrics formed in response to short bursts of sediment deposition followed by long periods of desiccation (Table 1; see Shinn 1983 for a detailed description of the tidal-flat environment).

Where wave energy dominates the shoreline environment, the tidal-flat environment is replaced by the beach environment. This commonly occurs in high-energy environments where grainstone shoals build out of water to form islands as well as along high-energy coastlines isolated from clastic sedimentation. The beach environment overlies supratidal sediments and is divided into foreshore and shoreface environments. The increase in current energy produced by the decrease in water depth as the shoreline is approached produces an upward-increase in grain size and sorting (see Inden and Moore 1983, for detailed description).

In arid climates, evaporite deposits may form by precipitation from standing bodies of marine water isolated from the ocean by tidal floats or grainstone bars (Lucia 1968, 1972; Lloyd et al. 1987). A hypersaline lagoon which is restricted from the ocean by a discontinuous barrier is the depositional model most commonly used to explain evaporite deposits. Sea water must be evaporated to about one third its original volume in order for gypsum to be deposited and to about one tenth of its volume in order for halite to be deposited. The amount of evaporation is a function of the net rate of evaporation, the amount of exchange between the hypersaline lagoon and the ocean, and the amount of hypersaline water that flows down in the underlying strata (reflux; Deffeyes et al. 1965). If there is no outflow, the salinity will be high and halite will be deposited; if there is a moderate amount of exchange, the salinity will be less and gypsum will be deposited; and if there is good exchange no evaporites will form.

Hypersaline lagoons vary in size and origin. The Gulf of Karabugas, which is part of the Caspian Sea, has an area of about 14,000 km^2 and is restricted from the Caspian Sea by a barrier bar. Coastal salinas are much smaller; the Pekelmeer on the island of Bonaire, Netherlands Antilles, has a surface are of about 3 km^2 and is isolated from the ocean by a coral beach. The types of restrictions vary as well. The Pekelmeer is fed by a flow of ocean water through the coral beach, whereas the Gulf of Karabugas is fed though a narrow channel connecting it to the Caspian Sea. In all known examples, however, the connection between hypersaline bodies of water and the open ocean is very small compared with the surface area of the hypersaline lagoon (Lucia 1972).

Depositional evaporites are recognized by laminated and coalesced-nodular textures. Laminated anhydrite suggests the deposition of gypsum crystals which are precipitated within the body of water and fall to the bottom as sediment. Coalesced-nodular textures suggest the growth of gypsum crystals attached to the bottom and later modified by diagenetic processes, such as the alteration from gypsum to anhydrite with burial (Warren and Kendall 1985). Beds of depositional evaporites are found immediately above subtidal deposits and within supratidal deposits. When anhydrite occurs within a subtidal sequence it most likely represents a change in sea level and marks the top of a high-frequency cycle. The occurrence of evaporites within supratidal deposits can suggest a change in sea level or a growing restriction related to sedimentary processes, such as the construction of a barrier bar.

Table 1. Sequence of sedimentary features in tidal-flat capped cycles (Lucia 1972)

Interpreted sedimentary environment	Sedimentary structures	Fossils	Texture
Supratidal	Irregular laminations Lithoclasts Mud Cracks Fenestre Pisolites Bedded anhydrite	Rare	Grain-dominated packstone to wackestone
Intertidal	Algal Stromatolites Burrowing	Few Gastropods, forams, ostracods	Packstone and wackestone
Tidal channel	Current lamination Cross-bedding	Few Echinoids, mollusks	Lithoclasts, fine sand-sized pellets and some mud
Subtidal	Burrowing	Locally abundant Typically a restricted fauna of gastropods, forams, and lamellibranchs	Grainstone to wackestone

Subtidal cycles deposited in the quiet waters of the middle-shelf environment are typically mud-dominated textures with an upward change in faunal content and increasing volume of allochems. As sedimentation fills accommodation space, however, the sediment surface is moved closer to wave and storm currents and grain-dominated packstones cap the cycles. Burrowing organisms churn the muddy sediment and produce pellets that, together with shell material, compose the grain fraction of the sediment. Local topography may produce sufficient current energy to produce grain-dominated fabrics and patch reefs. Whereas sediments deposited in this environment may have interpellet pore space, compaction of the pellets with burial commonly alters the fabric to a pelleted wackestone or mud-dominated packstone (see Enos 1983 and Wilson and Jordan 1983 for more detailed descriptions).

Grain-dominated packstones and grainstones generally are produced in the high-energy conditions found in the shelf crest by winnowing of lime mud (Figs.

9, 10). The classic upward shoaling succession of mud-dominated to ooid grain-stone textures typifies this environment (Fig. 11A; Ball 1967; Harris 1979). Typical high-energy deposits are (1) shelf margin sands where funneling of tidal energy by the shelf slope produces widespread tidal-bar and marine sand belts, (2) back-reef sands associated with landward transport of sediment from a rimmed-shelf fringing reef, and (3) local middle-shelf deposits associated with inner-island gaps or tidal inlets forming lobate tidal deltas.

Grain-dominated packstones are typically churned by burrowing organisms and do not show evidence of current transport. However, it is possible that some grain-dominated packstones result from the addition of carbonate mud by the mixing of muddy and grainstone sediments by burrowing. Cross bedding of all types and scales, from ripple bedding to large festoons, and commonly with multiple dip directions, indicating deposition out of tidal currents, is typically found in grainstone beds (see Halley et al. 1983 for detailed descriptions).

Reefs are concentrated at the high energy margins of rimmed shelfs where they can benefit from nutrients rising from the associated deep basins. The term "reef" has been much misused in the petroleum industry. At one time it was used as a general term for all carbonate reservoirs, and currently it is commonly used to describe any carbonate buildup. Here, the term "reef" will be used to describe carbonate bodies composed of bindstone or bafflestone because they may have unique pore structures related to the binding nature of the organisms. Reefs are the physical expression of a community of organisms that live and grow together and have little to do with physical depositional processes. Reefs, however, are a major source of carbonate sediment, and sedimentary deposits associated with reefs are usually grainstones because reefs are typically found in high energy environments. The organisms that form the reefs have evolved with time: Precambrian stromatolite reefs, Silurian coral reefs, Devonian stromat-oporoid reefs, Permian sponge reefs, and Cretaceous rudist reefs, to name a few. The result is a robust variety of reef textures and geometries.

A second type of carbonate accumulation that is commonly referred to as a reef is more correctly termed a carbonate buildup, mound, or bioherm. These are features with topographic relief that are formed in response to pre-existing topography or localized organic activity. They are composed of sediment with only minor bind- and bafflestone, and their petrophysical characteristics can be described in terms of depositional textures. They can start in shallow or deep water but tend to grow to sea level (see James 1983 for detailed descriptions of reefs and bioherms).

Subtidal cycles in low-energy environments found in slope and basin environments may be composed entirely of mud-dominated textures. The cycles may contain an upward increase in the percentage of allochems or they may contain graded beds and Bouma type turbidity sequences. HFCs are difficult to identify in these environments, and the cyclicity observed may not be related to the same process that produces cyclicity on the shelf.

Shelf slopes and basins adjacent to carbonate platforms receive carbonate sediment by mass transport from the shelf and by fallout of calcareous zoo-

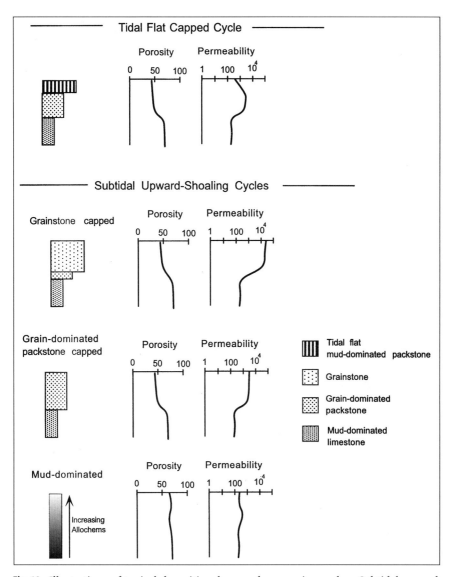

Fig. 12. Illustrations of typical depositional upward-coarsening cycles. Subtidal upward-shoaling cycles may be capped by grainstone, grain-dominated packstone, or mud-dominated packstone depending on the depositional environment. Tidal-flat capped cycles are most important because the tidal flat environment marks sea level. The vertical porosity and permeability profile is correlated with depositional texture

plankton and phytoplankton. The typically muddy sediment is punctuated with coarser grained sediment flows. Debris flows, slump deposits, reef talus, grain flows, and turbidites are typical transportation mechanisms. The type of sediment flow depends in part on the degree of slope. Breccia beds composed of reef talus and other debris flows are typically found associated with the steep slopes of a rimmed shelf, whereas slump deposits and grain flows are more typical of gentle carbonate ramp slopes. Turbidite deposits are found in lower slope and basin environments (see Enos and Moore 1983; Cook and Mullins 1983, Scholle et al. 1983 for more detail).

The porosity and permeability profiles in an HFC will reflect the vertical succession of depositional textures (Fig. 12). The base of the HFC is commonly a mud-dominated texture having 70% porosity and 200 md permeability. Grain-dominated packstone typically overlies mud-dominated strata and has 60 % porosity and 2000 md permeability. In high-energy environments, grainstone will cap the cycle with 45 % porosity and 30000 md permeability. The 2-D distribution of petrophysical properties is controlled by the lateral facies progression. The resulting 2-D pattern of petrophysical properties is shown in Fig. 9. Although there is a large spread in the permeability values, all the porosity and permeability values are sufficiently high to qualify as reservoir rock.

4.3.2
High-Frequency Sequence

Depositional textures are systematically stacked within high-frequency cycles and are related to current energy, topography, and organic activity. Assuming a constant rate of subsidence, each cycle begins with a flooding event produced by a sea-level rise, resulting in a transgressive, receding shoreline and an increase in space available for carbonate sediment to accumulate (referred to as accommodation space, Fig. 13). Sea-level rise is followed by a stillstand during which sediment partially or completely fills accommodation space. Tidal flat-capped successions indicate that all the accumulation space was filled and shoreline prograded seaward. The stillstand is followed by a sea level fall, reducing accommodation space, and resulting in a prograding, regressive shoreline and transport of sediment from the shelf into the basin (Fig. 13). During the sea level fall, the carbonate platform may be exposed, sedimentation may cease, and erosion may occur. The erosion may trigger debris slides into the basin, especially along steep shelf margins. A sea level rise produces another flooding event and the depositional cycle is repeated. *Flooding events approximate chronostratigraphic surfaces and define the HFC as a time-stratigraphic unit.*

The causes of repeated flooding events include cessation of carbonate sediment production due to filling of accommodation space (autocyclic), episodic subsidence, and externally forced fluctuations in sea level (allocyclic). Externally forced fluctuations are thought to be produced by changes in continental ice volume due to climatic changes, possibly related to systematic changes in the earth's orbit (eustacy). At present, eustacy is the popular hypothesis, and it is be-

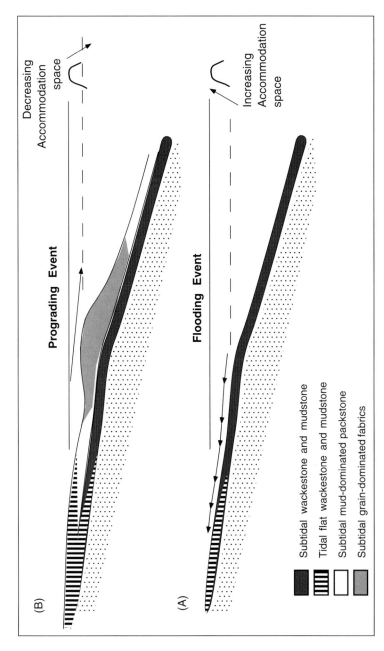

Fig. 13. Diagram illustrating the development of a depositional cycle through a combination of structural subsidence and eustatic sealevel fluctuations. Flood of the shelf occurs with sea-level rise which increases accommodation space for sediment to fill. Progradation occurs when sealevel falls. Decreasing accommodation space results in sediments building to sea-level, higher energy environments, and transport of sediment on and off the shelf

lieved that the cyclicity is forced by glacio-eustatic cycles. It is well known that during the Pleistocene, the sea level fluctuated in response to the waxing and waning of continental glaciers. For our purposes, the origin of the HFC is not an issue, and we will assume a eustatic origin for the cyclicity commonly observed in carbonate reservoirs.

Repeated eustatic sea-level cycles result in the vertical stacking of HFC (Fig. 14). Cycles are stacked vertically into retrogradational cycles, aggradational cycles, and progradation cycles. Retrogradational cycles are formed when the eustatic sea-level fall for each cycle is much less than the rise. The shoreline will move farther landward with each successive cycle, a pattern described as back-stepping or transgression. The sediments are said to be deposited in the transgressive systems tract (TST). Aggradational cycles are formed when the eustatic rise and fall is equal, and the resulting facies will stack vertically. These cycles are defined as part of the highstand system tract (HST). Progradational cycles form when the eustatic fall for each cycle is greater than the rise. The shoreline for each successive cycle will move seaward, a pattern described as progradation or regression, and the sediments are said to be deposited in the highstand systems tract (HST). Sediments deposited when relative sea level is lowest are said to be deposited in the lowstand systems tract (LST). The sequence from TST to HST to LST defines a larger sea level signal referred to as a high-frequency sequence (HFS; fig 14).

Cycles deposited in the transgressive systems tract will typically be subtidal cycles with rare tidal-flat-capped cycles. Textures may coarsen upward, becoming grain-dominated in association with topographic slope changes. Tidal flats will be restricted to the shoreline and will not prograde far seaward because of the overall transgressive nature of the shoreline. The overall sea level rise results in each successive cycle starting at a more landward position as shown by cycles 1, 2, and 3 in Fig. 14. Cycles deposited in the highstand systems tract are higher energy deposits because the amount of accommodation space is less due to more sea level fall than rise. They typically have well-developed tidal-flat-capped cycles landward, lagoonal mud-dominated subtidal cycles, grain-dominated packstone and grainstone capped cycles and reefs at the shelf crest, and mud-dominated cycles in the outer shelf and basinal positions (see cycles 4, 5, and 6 in Fig. 14). Because sea level fall is larger than rise in the HST, the cycles begin to onlap the shelf. Sedimentation fills accommodation space so that (1) tidal flats have ample opportunity to prograde seaward and, (2) shelf sediment is transported basinward, producing outer-shelf clinoform deposits.

This process produces a systematic pattern of depositional textures organized within the high-frequency sequence illustrated in Fig. 15. Petrophysical properties will be distributed according to the rock-fabric facies patterns with the highest permeability in the grainstones and the lowest permeability in the carbonate muds. The resulting distribution of petrophysical properties is shown in Fig. 16. The result is a time-transgressive belt of highly permeable grain-dominated sediment bounded basinward by low-permeability mud-dominated slope and basin sediments, and landward by low-permeability mud-dominated middle-shelf

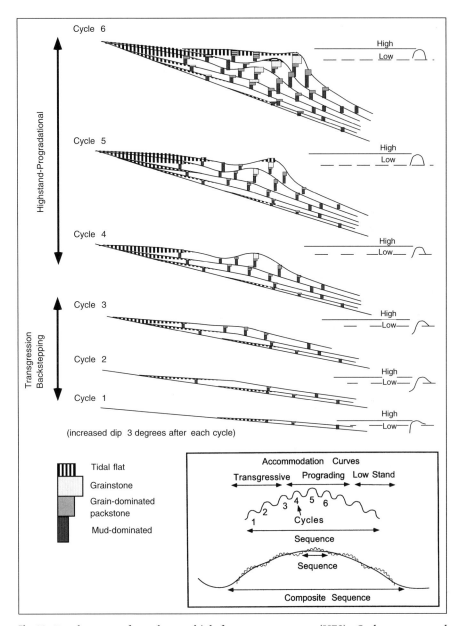

Fig. 14. Development of a carbonate high-frequency sequence (HFS). Cycles are grouped into longer term sealevel signals referred to as an HFS. Cycles backstep when long-term sealevel rise is greater than the long-term fall. Cycles prograde when long-term fall is greater than long-term rise. HFSs can be grouped into a longer-term sealevel signal referred to as a composite sequence

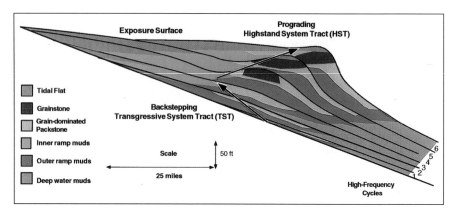

Fig. 15. Diagram of the HFS constructed in Fig. 13 showing the distribution of depositional textures and high-frequency cycles. Grainstones are concentrated in the ramp crest facies tract of the highstand systems tract

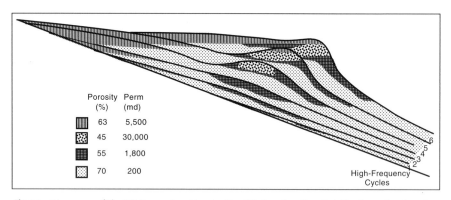

Fig. 16. Diagram of the HFS construction in Fig. 13 showing the distribution of petrophysical properties based on depositional textures. Highest permeability is concentrated in the ramp crest location and in the tidal flats

sediment. The value of the HFC is to demonstrate the time-transgressive nature of the high-permeability grain-dominated facies and illustrate that these facies are separated by layers of low-permeability mud-dominated sediment.

4.4
Example

An excellent example of the systematic distribution of rock-fabric facies in a carbonate sequence is found in the San Andres Formation that outcrops on the Algerita Escarpment, Guadalupe Mountains, Texas and New Mexico (Fig. 17 ; Kerans et al. 1994). Regional mapping shows that the 1500-ft-thick San Andres Formation can be divided into multiple high-frequency and composite sequences. Each sequence is composed of high-frequency cycles, and the facies patterns

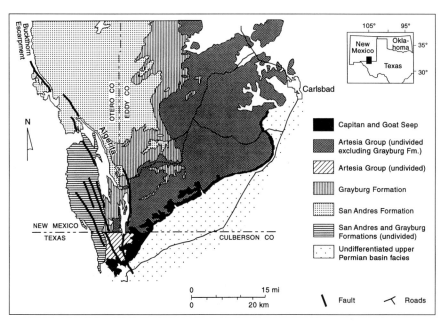

Fig. 17. Geological map of the Guadalupe Mountains area showing the location of the Algerita Escarpment San Andres outcrops. (Kerans et al., 1994)

within the cycles and facies offset between cycles define the TST and HST. Apparently, the peritidal environment has been eroded, leaving the middle ramp, ramp crest, outer ramp, and ramp slope and basin exposed for study. Petrophysical data collected from the outcrop shows dolograinstones with 20 md permeability, grain-dominated dolopackstones with 4 md permeability, dolowackestones with 0.4 md permeability, and dolomudstones with less than 0.1 md permeability.

Detailed mapping was done on a reservoir scale window within the formation that is a close analog for many San Andres reservoirs of the Permian Basin. The reservoir window is about 1 mile (1.6 km) long and 150 ft (30 m) high. Facies descriptions from numerous measured sections indicate nine high-frequency cycles that can be mapped along the length of the outcrop (Fig. 18). The cycles are predominantly asymmetric upward-shallowing cycles, and typically begin with basal mudstone, followed by wackestone or mud-dominated packstone, grain-dominated packstone, grainstone, and locally a fenestral (tidal flat) cap. In Fig. 18, cycles 3 and 8 have tidal-flat caps; cycles 1, 2, 3, 7, and 9 are subtidal cycles with thick grainstone or grain-dominated packstone caps; and cycles 4, 5, and 6 have thin grain-dominated packstone caps. This vertical succession suggests a sequence boundary at the contact between cycles 3 and 4, and between cycle 9 and overlying cycles that onlap the cycle 9 exposure surface. Cycles 1-3

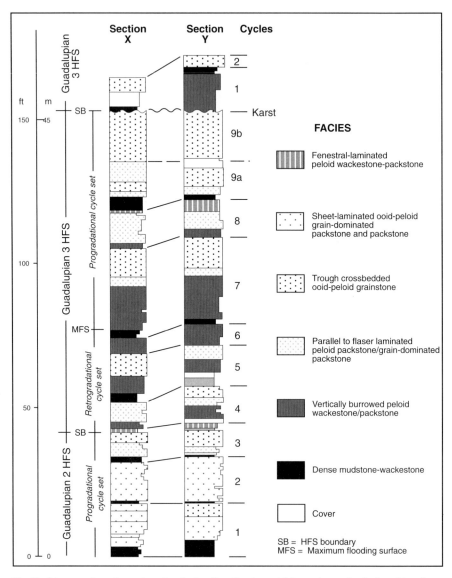

Fig. 18. Measured sections showing facies distribution within cycles and relationship of cycles to HFS framework in the Lawyer Canyon reservoir window. (Kerans et al. 1994)

are prograding cycles of a lower highstand systems tract and cycles 7-9 are prograding cycles of an upper highstand systems tract. Cycles 4-6 are retrogradational cycles of the transgressive systems tract. These two sequence boundaries have been traced throughout the Guadalupe Mountains and formalized as Guadalupian 2 high-frequency-sequence, and Guadalupian 3 high-frequency-sequence by Kerans et al. (1994) and Kerans and Fitchen (1995).

The measured sections provide one-dimensional data similar to a vertical development well. In excellent outcrops, however, two-dimensional data can be obtained by mapping the facies laterally between sections on the ground or by using oblique aerial photography. The cycle boundaries described here have been traced for 1 mile and the depositional facies mapped. The facies change laterally in a gradual manner and without sharp contacts, and they do not cross cycle boundaries. The vertical succession of facies in the middle ramp are typically more complicated than those of the ramp crest, and cycles of the outer ramp are simple upward-coarsening cycles from mudstone and wackestone to mud- or grain-dominated packstones (Fig. 19). By mapping laterally, the vertical succession of facies within a cycle can be converted into a two-dimensional facies progression model (Fig. 19). Fossils that are water depth indicators are important in interpreting the lateral facies progression, and fusulinids are key water depth indicators for the San Andres. As shown on the cross section, fusulinids are concentrated in the outer ramp and distal outer ramp. Cherty mudstones are also characteristic of the distal outer ramp.

Grainstones are of particular interest because they are the most permeable facies in modern sediments and in this outcrop. The detailed facies maps from the reservoir window show a partitioning of grainstone bodies between the

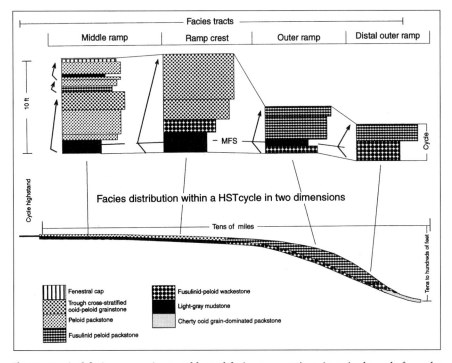

Fig. 19. Vertical facies successions and lateral facies progressions in a single cycle from the San Andres outcrop. (Kerans and Fitchen 1995)

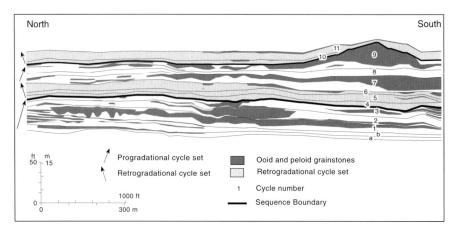

Fig. 20. Distribution of grainstone facies in the Lawyer Canyon, Algerita Canyon reservoir window showing grainstones concentrated in progradational cycle sets in two HFSs. (Kerans et al. 1994)

transgressive and highstand systems tracts at the HFS scale (Figs. 20 and 21 Kerans et al. 1994). Grainstones are more common and have greater dimensions in the highstand than in the transgressive systems tracts. Also, grainstones are more common in the ramp-crest facies tract than in the inner-ramp or outer-ramp tracts.

This outcrop example illustrates the systematic distribution of depositional textures in a carbonate platform. Assuming the products of diagenesis conform reasonably well to depositional textures, predictions of petrophysical properties can be made based on predictions of the three-dimensional pattern of petrophysically significant depositional facies. Depositional patterns in carbonates are highly variable, and we have not attempted to discuss the extensive literature available on the subject. Instead, we have presented a simplified version of carbonate sedimentation and cycle stacking. However, this approach can be applied to many major carbonate reservoirs in the world.

4.5
Summary

Carbonate sediments have a wide range of particle size and sorting because they are formed by organic activity and because current transport is not a major factor in the formation of depositional textures. The petrophysical properties are highly variable; however, pore-size distribution is always a function of the particle type, size, and sorting. Porosity values range from 40 to 75% and permeabilities from 200 to 30000 md. Mud-dominated fabrics average 70% porosity and 200 md permeability, grain-dominated packstones average 55% porosity and 1800 md permeability, and grainstones average 45% porosity and 30000 md permeability.

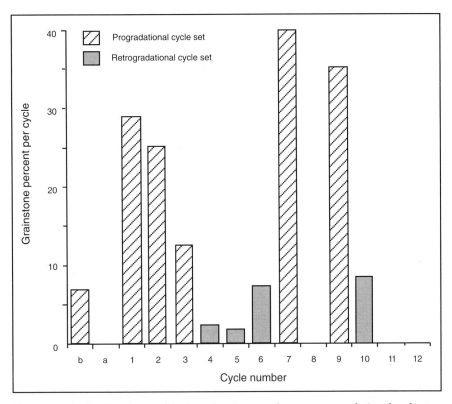

Fig. 21. Graph showing the partitioning of grainstones between progradational and retrogradational cycles. San Andres Formation, Algerita Escarpment. (Kerans et al. 1994)

The spatial distribution of petrophysical properties is linked to facies patterns. Rock-fabric facies are systematically distributed within high-frequency cycles and within high-frequency sequences. These are chronostratigraphic units bounded by time surfaces that can be correlated from well to well. Depositional textures are vertically stacked into tidal-flat capped cycles and subtidal cycles which may be capped by bindstone, grainstone, grain-dominated packstone, mud-dominated packstone, or wackestone depending on the depositional energy. Depositional energy is controlled by topography and the types of ocean currents. The highest energy is generally located at the shelf margin and the cycles are typically capped by bindstone, grainstone, or grain-dominated packstone. Gentle currents are typically found over the middle shelf, except during storms, and mud-dominated cycles with thin mud- to grain-dominated packstones are cycle caps. The shoreline is a trap for sediment transported from the subtidal to the shore, forming beaches and tidal-flat-capped cycles. A variety of currents are found basinward of the shelf crest, depositing mud-dominated sediment as well as graded beds and boulder beds. The two basic cycles, however, are tidal-flat capped and subtidal, and subtidal cycles are commonly composed of two basic textures, a lower mud-dominated texture and an upper grain-dominated cap.

Each HFC begins with a flooding event produced by a relative sea level rise. Flooding events approximate chronostratigraphic surfaces and define the HFC as a time-stratigraphic unit. High-frequency cycles stacked into retrogradational cycles indicate an overall sea level rise, aggradational cycles indicate a sea level still stand, and progradational cycles indicate a general level fall. The sequence from retrogradational to progradational defines a larger sea-level signal and is referred to as a high-frequency sequence. The systematic patterns of depositional textures organized within the high-frequency sequence define the distribution of petrophysical properties at the cycle scale.

There are no nonproductive areas in the depositional model because very few sediments can be considered nonreservoir quality. However, bodies of high-permeability sediment are located in the vicinity of the shelf crest and are bounded seaward and landward by low-permeability mud-dominated sediments. Petroleum reservoirs commonly have nonproductive areas because diagenetic processes modify depositional texture, most commonly reducing porosity and permeability.

Carbonate sedimentary textures are systematically distributed on a carbonate platform. Assuming the products of diagenesis conform reasonably well to the depositional textures, predictions of petrophysical properties can be made based on predictions of the three-dimensional patterns of petrophysically significant depositional facies. Depositional patterns in carbonates are highly variable, and there is an extensive literature available on the subject.

In the next chapters we will examine diagenetic processes and their effect on petrophysical properties as the depositional textures are converted to rock-fabrics. Importantly, we will focus on the problem of prediction in the presence of various diagenetic processes.

References

Ball MM (1967) Carbonate sand bodies of Florida and the Bahamas. J Sediment Petrol 37: 556–591

Cook HE, Mullins HT (1983) Basin margin environment. In: Scholle PA, Bebout D, Moore HM (eds) Carbonate depositional environments. AAPG Mem 33: 539–618

Deffeyes KS, Lucia FJ, Weyl PK (1965) Dolomitization of Recent and Plio-Pleistocene sediment by marine evaporate waters on Bonaire, Netherlands Antilles. In: Pray LC, Murray RC (eds) Dolomitization and limestone diagenesis – a symposium. SEPM Spec Publ 13: 71–88

Dunham RJ (1962) Classification of carbonate rocks according to depositional texture. In: Ham WE (ed) Classifications of carbonate rocks – a Symposium. AAPG Mem 1: 108–121

Embry AF, Klovan FE (1971) A late Devonian reef tract of northeastern Banks Island, N.W.T. Bull Can Pet Geol 19: 730–781

Enos P (1983) Shelf environment. In: Scholle PA, Bebout D, Moore HM (eds) Carbonate depositional environments. AAPG Mem 33: 267–296

Enos P, Moore CH (1983) Fore-reef slope environment. In: Scholle PA, Bebout D, Moore HM (eds) Carbonate depositional environments. AAPG Mem 33: 507–538

Enos P, Sawatsky LH (1981) Pore networks in Holocene carbonate sediments. J Sediment Petrol 51, 3: 961–985

Gebelein CD, Steinen RP, Garrett P, Hoffman EJ, Queen JM, Plummer LN (1980) Subsurface dolomitization beneath the tidal flats of central West Andros Island, Bahamas. In: Zenger DJ, Duhnam JB, Ethington RL (eds) Concepts and models of dolomitization. SEPM Spec Publ 28: 31–49

Gerhard LC (1985) Porosity development in Mississippian pisolitic limestone of the Mission Canyon Formation, Glenburn Field, Williston Basin, North Dakota. In: Roehl PO, Choquette PW (eds) Carbonate petroleum reservoirs. Springer, Berlin Heidelberg New York, pp 193–205

Halley RB, Harris PM, Hine AC (1983) Bank margin environment. In: Scholle PA, Bebout D, Moore HM (eds) Carbonate depositional environments. AAPG Mem 33: 463–506

Harris PM (1979) Facies anatomy and diagenesis of a Bahamian ooid shoal. University of Miami, Florida, Comparative Sedimentology Laboratory, Sedimenta 7, 163 pp

Inden RF, Moore CH (1983) Beach environment. In: Scholle PA, Bebout D, Moore HM (eds) Carbonate depositional environments. AAPG Mem 33: 211–267

James NP (1983) Reef environment. In: Scholle PA, Bebout D, Moore HM (eds) Carbonate depositional environments. AAPG Mem 33: 345–462

James NP (1984) Reefs. In: Walker RG (ed) Facies models, 2nd edn. Geoscience Canada, Reprint Series 1, Geological Association of Canada, Ottawa, pp 229–244

Kerans C, Fitchen WM (1995) Sequence hierarchy and facies architecture of a carbonate-ramp system: San Andres Formation of Algerita Escarpment and western Guadalupe Mountains, West Texas and New Mexico. The University of Texas at Austin, Bureau of Economic Geology, Report of Investigation 235, 85 pp

Kerans C, Lucia FJ, Senger RK (1994) Integrated characterization of carbonate ramp reservoirs using outcrop analogs. AAPG Bull 78, 2: 181–216

Lloyd RM, Perkins RD, Kerr SD (1987) Beach and shoreface ooid deposition on shallow interior banks, Turks and Caicos islands, British West Indies. J Sediment Petrol 57, 6: 976–982

Lucia FJ (1968) Recent sediments and diagenesis of south Bonaire, Netherlands Antilles. J Sediment Petrol 38, 3: 845–858

Lucia FJ (1972) Recognition of evaporate-carbonate shoreline sedimentation. In: Rigby JK, Hamblin WK (eds) Recognition of ancient sedimentary environments. SEPM Spec Publ 16: 160–191

Milliman JD, Freile D, Steiner RP, Wilber RJ (1993) Great Bahama Bank aragonite mud: mostly inorganically precipitated, mostly exported. J Sediment Petrol 63, 4: 589–695

Patterson RJ, Kinsman DJJ (1981) Hydrologic framework of a sabkha along Arabian Gulf. AAPG Bull 65, 8: 1457–1475

Reed JF (1985) Carbonate platform facies models. AAPG Bull 69, 1: 1–21

Schmoker JW, Halley RB (1982) Carbonate porosity versus depth: a predictable relation for south Florida. AAPG Bull 66, 12: 2561–2570

Schmoker JW, Krystinic KB, Halley RB (1985) Selected characteristics of limestone and dolomite reservoirs in the United States. AAPG Bull 69, 5: 733–741

Scholle PA, Arthus MA, Ekdale AA (1983) Pelagic environment. In: Scholle PA, Bebout D, Moore HM (eds) Carbonate depositional environments. AAPG Mem 33: 619–692

Shinn EA (1983) Tidal flat environment: In: Scholle PA, Bebout D, Moore HM (eds) Carbonate depositional environments. AAPG Mem 33: 172–210

Walker RG (1984) Facies models, 2nd edn. Geoscience Canada, Reprint Series 1, Geological Association of Canada, Ottawa, 317 pp

Walter LM (1985) Relative reactivity of skeletal carbonates during dissolution: implication for diagenesis. In: Schneidermann N, Harris PM (eds) Carbonate cements. SEPM Spec Publ 36: 3–16

Warren FK, Kendall CGSC (1985) Comparison of sequences formed in marine Sabkha (subaerial) and salina (subaqueous) settings – modern and ancient. AAPG Bull 69, 6: 1013–1023

Wilkinson BH (1979) Biomineralization, paleoecology and the evolution of calcareous marine organisms. Geology 7: 524–527

Wilson JL, Jordan C (1983) Middle shelf environment. In: Scholle PA, Bebout D, Moore HM (eds) Carbonate depositional environments. AAPG Mem 33: 267–344

Diagenetic Overprinting and Rock Fabric Distribution: The Cementation, Compaction, and Selective Dissolution Environment

5.1
Introduction

The three-dimensional spatial distribution of petrophysical properties is controlled by the spatial distribution of geologic processes, processes that can be separated into depositional and diagenetic. In Chapter 4 we discussed depositional processes and focused on (1) the origin of depositional textures, (2) the relationship between porosity, permeability, and depositional texture, (3) the vertical and lateral distribution of depositional textures related to topography, current energy, biologic activity, and eustatical controlled cyclicity, and (4) the fundamentals of sequence stratigraphy. The importance of chronostratigraphic surfaces was emphasized as the fundamental element in constructing a geological framework within which petrophysically-significant depositional textures can be systematically distributed.

Whereas it is clear that the three-dimensional spatial distribution of petrophysical properties is initially controlled by patterns of depositional textures, it is also clear from reservoir studies that the petrophysical properties found in carbonate reservoirs are significantly different from those of modern carbonate sediments. Diagenesis typically reduces porosity, redistributes the pore space, and alters permeability and capillary characteristics. The porosity of modern sediments ranges from 40 to 70 % whereas the porosity of carbonate reservoirs in the USA ranges from 9 to 17% (Schmoker et al. 1985), and the permeability of carbonate reservoirs is similarly reduced over that found in carbonate sediments. Therefore, an understanding of diagenetic processes and the patterns of their products is essential for carbonate reservoir description and the construction of reservoir models.

The basic diagenetic processes discussed here are (1) calcite carbonate cementation, (2) mechanical and chemical compaction, (3) selective dissolution, (4) dolomitization, (5) evaporite mineralization, and (6) massive dissolution, cavern collapse, and fracturing. Each of these processes can be identified by a specific fabric. Occlusion of pore space by calcite can be identified by crystal geometry and position relative to grain fabrics. Compaction can be identified by grain interpenetration, grain breaking, grain deformation, grain spacing, and stylolitization. Selective dissolution refers to the formation of pore space by the removal of a specific rock-fabric element forming a fabric-selective vuggy pore type. Massive dissolution refers to the formation of relatively large pore space without regard to rock fabric, and the voids may be sufficiently large to collapse forming collapse breccia and associated fracture patterns. Dolomitization is identified by the presence of the mineral dolomite and evaporite mineralization is identified by the presence of evaporite minerals such as anhydrite and gypsum.

Whereas each of these diagenetic processes can be identified and studied independently, they overlap in time and place and therefore have an effect on each other. A sedimentary deposit is formed only once whereas it may be overprinted by any or all of the diagenetic processes listed above. Petrographic studies are useful for determining the sequence of diagenetic events: these studies often show a great deal of overlap in diagenetic timing, and may show that processes recur. Indeed, diagenesis is a continuing process, starting with the cessation of sedimentation and ending with the onset of metamorphosis. Thus, the sequence of diagenetic events may be extremely complicated and the pattern of diagenetic products difficult to predict if they are not related to depositional patterns.

The distribution of diagenetic products is controlled by the nature of the precursor carbonate, and with increasing time the nature of the precursor carbonate becomes more and more unlike depositional textures. Compaction starts with burial and may end only with uplift and exposure. Cementation can be penecontemporaneous with deposition or later during burial, and early cementation may inhibit later burial compaction. Dolomitization can occur immediately after sedimentation or millions of years after deposition; and the cementation, dissolution, and compaction that occurs before dolomitization will affect the pattern of dolostone. Early dolomitization may produce stratiform dolostone whereas late dolomitization may be controlled by fractures and collapse breccias and form discontinuous bodies of dolostone.

Therefore, to distribute petrophysical properties within a carbonate reservoir model, the process of diagenetic overprinting of depositional textures must be understood. A key question is the degree of conformance between diagenetic products and depositional patterns. If transport of material in and out of the system is not an important factor in producing the diagenetic product, the product will generally conform to depositional patterns. However, if transport of ions in and out of the system by fluid flow is required to produce the diagenetic product, then the diagenetic product may not conform to depositional patterns. In this case, knowledge of the geochemical-hydrological system may be required to map the diagenetic products, including knowing the source of the fluid and the direction of fluid flow.

Diagenetic processes are grouped according to their conformance to depositional patterns because of our focus on mapping and predicting the distribution of petrophysical rock-fabrics. Cementation, compaction, and selective dissolution form the first group; reflux dolomitization and evaporite mineralization form the second group; and massive dissolution, collapse brecciation, and fracturing make up the third group.

The products of cementation, compaction, and selective dissolution can normally be linked to depositional textures. Compaction and associated cementation is a diagenetic process that is a function of rock strength and the time-overburden history (Weyl 1959). Selective dissolution of unstable aragonite allochems and associated precipitation of calcite cement may occur without the long distance transport of ions with or without the introduction of a local meteoric lens (Budd and Land 1990). Early cementation processes require fluid flow to import calcium and carbonate into the system, but the fluid flow is closely tied to the depositional environment through permeability. Late or burial cementation may occur by chemical compaction and thus be linked to a depositional environment, or it may require regional transport of ions by ground water and not be linked to a depositional environment.

Reflux dolomitization and evaporite mineralization form the second group. Dolomitization requires fluid flow for the introduction of magnesium into the system. Therefore, fluid flow is an important element in the origin of dolomite fabrics, and dolomite patterns may not be linked to depositional patterns. Dolomitization may increase the particle size significantly, modifying the pore-size distribution of the sediment and smoothing out important petrophysical differences found in depositional textures. Predolomite diagenetic history may significantly alter the permeability structure and result in dolomitizing waters following diagenetic rather than depositional flow paths. Gypsum and anhydrite are commonly associated with dolomitization and require the transport of sulfate into the system by hypersaline water. Studies have shown little linkage between depositional facies and patterns of diagenetic gypsum or anhydrite. However, sulfate commonly selectively occludes pore space in grainstones while forming poikilotopic anhydrite in other fabrics.

Massive dissolution, collapse brecciation, and fracturing make up the third group, and their products are controlled by precursor diagenetic events and ground water flow. Nonfabric selective, massive dissolution clearly results in the reorganization of pore space through the removal of carbonate from some areas and reprecipitation in other areas through a complex geochemical-hydrological meteoric flow system. Massive dissolution processes significantly modify rock fabrics and create touching-vug pore systems that have little relationship to depositional patterns.

This chapter will focus on the cementation, compaction, and selective-dissolution diagenetic environment. The remaining two environments will be addressed in the following chapters.

5.2
Cementation/Compaction/Selective Dissolution

The processes of cementation, compaction, and selective dissolution comprise a diagenetic environment characterized by shallow marine diagenesis, shallow meteoric diagenesis related to local subaerial exposure of shoals and tidal flats, and simple burial in sea water and ground water. Cementation and compaction reduce depositional porosity and systematically reduce pore-size. Selective dissolution typically forms separate-vug porosity by selectively dissolving grains composed of unstable minerals. The dissolved carbonate may be precipitated in the form of calcite cement in adjacent pore space. The patterns of these diagenetic products are normally closely linked to depositional textures and can be predicted by mapping the latter.

5.2.1
Calcium Carbonate Cementation

Calcium carbonate cementation occludes pore space and reduces pore size. Studies have shown typical textures to be isopachous fibrous and bladed cements, equant or blocky sparry cements, syntaxial cements, pendulous cements, meniscus cements, and radiaxial cements (Harris et al. 1985; Hurley and Lohmann 1989; Fig. 1). Calcite cements are composed of calcite, high-magnesium calcite, or aragonite when they are formed. However, high-magnesium calcite and aragonite are unstable minerals and are replaced by the stable form, calcite, with time and burial.

Cementation starts soon after deposition. Indeed, clasts of cemented sediment, called intraclasts, are found redeposited in later sediment. Early cementation of sediments in the shallow burial environment results from the circulation of large quantities of marine water through very permeable sediments such as grainstones and reef debris (Shinn 1969; James and Ginsburg 1979). The driving force is tidal and wave energy characteristic of high-energy environments occupied by reefs and grainstone sediments. Marine cement typically precipitates evenly around the grains as fibrous or bladed cement and is referred to as isopachous cement. Large voids commonly found in carbonate reefs are often filled with large botryoidal fans of radiaxial marine carbonate cement (Fig. 1).

Cementation by calcite continues as the sediment is buried (Heydari and Moore 1993). The process of burial cement is poorly understood. Budd et al. (1993) report no burial cementation in the Florida Aquifer. Instead, cementation is precompaction and related to the Oligocene exposure surface or early mineralogical stabilization. Heydari and Moore (1989, 1993), however, describe burial cementation related to thermochemical sulfate reduction. The source of the calcium and carbonate ions required for continued cementation are (1) grain dissolution associated with chemical compaction, (2) dissolution of unstable minerals such as aragonite, and (3) long distance transport of ions by ground-water flow. Common burial cement fabrics are equant calcite spar and syntaxial overgrowths. Equant cements are often unevenly distributed throughout the rock, producing a small-scale heterogeneity that may explain some small-scale permeability variability. Syntaxial cements are overgrowths on grains that are composed of a single calcite crystal. Echinoderm fragments commonly have syntaxial overgrowths, and pore space in sediments composed predominantly of echinoderm particles is commonly totally occluded (Lucia 1962).

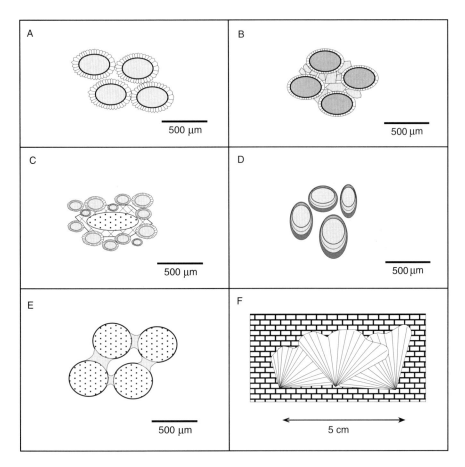

Fig. 1. Common calcite cement textures. **A** Fibrous or bladed isopachous cement: **B** equant or blocky sparry cement, **C** syntaxial or overgrowth cement: **D** pendular or microstalactitic cement: **E** meniscus cement: **F** radiaxial or botryoidal cement

Several cement fabrics are unique to the vadose zone (zone of air saturation). Meniscus cement is unique because is retains the shape of the air/water interface (Fig. 1). It is confined to pore throats and reduces permeability more than would be expected from a similar volume of isopachous or equant cement. Pendulous or microstalactitic cements preferentially grow down from the bottom of grains, suggesting growth from waters percolating downward through partially air-filled pores.

A characteristic of all cements is that they propagate from pore walls into pore space and therefore reduce pore size as they grow. In the case of evenly distributed cement, pore size is reduced in proportion to the amount of cement precipitated (Fig. 2). The systematic reduction in pore size accounts for the observed systematic changes in permeability and capillary properties with chang-

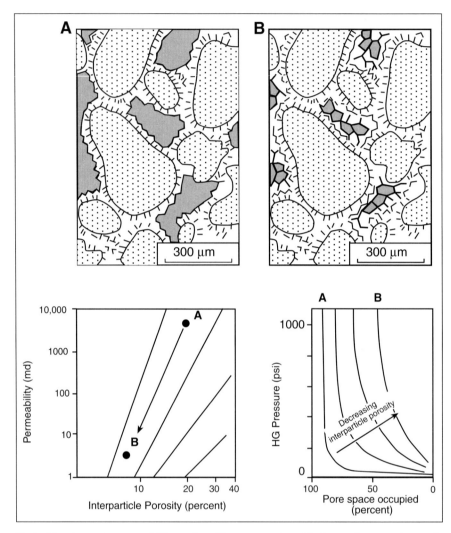

Fig. 2. The change in permeability and capillary properties with decreasing porosity resulting from occlusion of intergrain pore space. As 20% intergrain pore space in sample A is reduced by cementation to 7% in sample B, permeability is systematically reduced and the
shape of the capillary pressure curve is systematically changed to reflect smaller pore sizes

es in interparticle porosity. Therefore, pore-size distribution is a function of interparticle porosity, grain size, and sorting (Fig. 3).

Unevenly distributed calcite cement will produce small-scale heterogeneity. The
total porosity of the sample will be reduced, but the porosity and pore size of the
noncemented portions will remain unchanged. Because permeability and capillary

Fig. 3. Cross plot of log permeability and interparticle porosity for nonvuggy limestones showing rock-fabric fields. The scattering of data within a field is due to the variability of grain size, sorting, and cement distribution and size

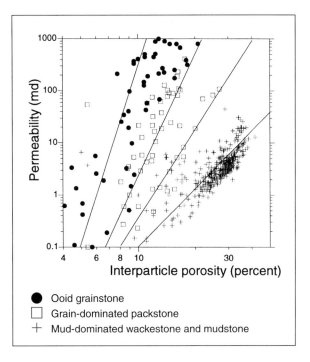

properties are principally a function of pore size, they will be reduced less than they would be if the same volume of cement were evenly distributed. Therefore, patchy calcite cement reduces porosity but may not have a significant effect on permeability and capillary properties.

5.2.2
Compaction

Compaction effects are difficult to separate from cementation effects, but they both reduce pore-size and porosity. Compaction is both a physical and chemical process resulting from the increased overburden pressure due to burial. Textural effects include the loss of porosity, reduction of pore-size, grain penetration, grain breaking, grain deformation, and microstylolites. It does not require the addition of material from an outside source, and is a function of texture only. In addition, compaction is a source of energy to move fluids out of the sediment and into adjacent sediments, usually flowing upward.

Experimental data has shown that simple mechanical compaction can reduce the porosity of lime muds from 70% to about 40% within the first 100 meters (330 ft) of burial (Fig. 4) (Goldhammer 1997). Soft fecal pellets will be compacted, changing the texture from a pellet grain-dominated packstone to a pellet mud-dominated packstone or wackestone fabric. Hard pellets will retain their shape and intergrain pore space. In contrast, grain-supported sediments can re-

Fig. 4. The change in porosity of mechanically compacted mud and sand sediments with depth (Goldhammer 1997). The potential effect of chemical compaction and cementation is shown by the difference between mechanical compaction and subsurface porosity vs. depth data from South Florida (Schmoker and Halley 1982)

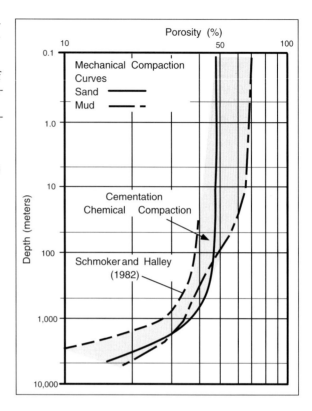

tain the original porosity of 47% to a depth of about 700 m (2310 ft) before closer packing and grain breakage begin to significantly reduce pore space (Fig. 4).

In addition to porosity loss through mechanical compaction, chemical compaction in the form of pressure solution at grain contacts will also result in porosity reduction with burial and time. Pressure solution of carbonate grains provides calcium and carbonate ions that can be precipitated as cement in adjacent pore space. Schmoker and Halley (1982) for South Florida present an example of the combined effect of cementation and compaction with time and burial. The comparison of this curve with the mechanical compaction curve illustrates that cementation and chemical compaction are important processes in the loss of porosity with burial (Fig. 4).

The compaction curves suggest that carbonate sediments lose their porosity slowly with burial. Porosity values between 30 and 40% should not be unusual at depths of 1000 m (3000 ft). This has been observed in the Neogene of the Bahama Platform, where core and log data demonstrate consistent high porosity values of 35 to 50% at depths of 700 m (2100 ft) (Anselmetti and Eberli 1993).

5.2.3
Selective Dissolution

Dissolution is the diagenetic process by which carbonate and evaporite minerals are dissolved and removed, thus creating and modifying pore space in reservoir rocks. The effect of this process on permeability depends upon the geometry and location of the resulting voids relative to the rock fabric. Dissolution can be fabric selective and form moldic pores, referred to here as separate vugs. In other cases, dissolution is not fabric selective and results in interconnected voids referred to here as touching vugs. This type of dissolution will be discussed in Chapter 7.

Selective dissolution occurs when one fabric element is selectively dissolved in preference to others. This usually results from the fact that carbonate sediments are composed of minerals with different solubility. Anhydrite and gypsum are more soluble than calcite or dolomite and are commonly dissolved selectively to form sulfate molds. Laboratory experiments have shown that solubility of calcium carbonate minerals increases from low-magnesium calcite (LMC) to aragonite to high-magnesium calcite (HMC, see chap. 4). However, depositional particles originally composed of high-magnesium calcite are commonly replaced with low magnesium calcite rather than dissolved. The common observation in the geological record is that aragonite grains tend to be dissolved in preference to LMC and HMC (Swirydezuk 1988; Melim et al. 1998).

Formation of vugs by selective dissolution commonly inverts pore-space fabrics in grainstones from intergrain porosity and solid grains to moldic porosity and occluded intergrain pore space. Comparative data suggest that porosity does not increase as a result of this process (Fig. 5). Modern ooid grainstones have 45% porosity, and ooid grainstones with moldic pore space rarely have more than 30% porosity. Core analyses of moldic grain/packstone samples taken from wells Unda and Clino, from below the upper karsted interval all show porosity values between 40 and 50% (Melim et al. 1998), suggesting only a slight reduction in porosity from the time of deposition. Most likely, the porosity does not increase because the calcium and carbonate ions produced by dissolution of unstable aragonite are precipitated as calcite cement in the juxtaposed pore space. This can occur with burial in the presence of marine ground water (Dix and Mullins 1988) or in the shallow burial environment in the presence of a localized freshwater lens such as those found in grainstone shoals (Budd and Land 1990).

Moldic pore space is connected only through the interparticle pore network. As a result, moldic pore space contributes little to permeability. The presence of separate vugs in limestones and dolostones results in less permeability than would be anticipated if all the porosity were interparticle (Lucia 1983).

Enlargement of interparticle pore space is another type of selective dissolution. The resulting enlargement of interparticle pore space improves flow characteristics and capillary properties, the reverse of moldic pore space. For example, intergrain pore space in some Jurassic grainstone reservoirs in the Gulf Coast, USA has been enlarged by dissolution in the presence of late subsurface

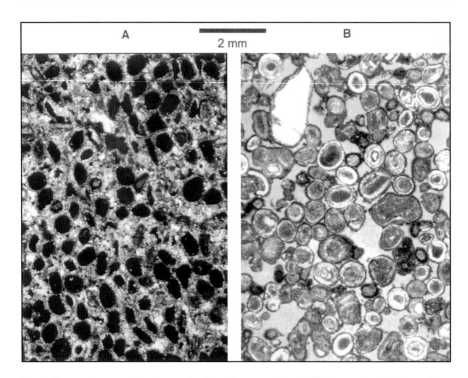

Fig. 5. Inverted texture. Both samples have between 20 and 25% porosity. A Photomicrograph of moldic porosity (black) with occluded intergrain pore space. B Photomicrograph of a grainstone with intergrain porosity

brines containing high concentrations of H_2S (Moore and Druckman 1981). In another example, intergrain lime mud in a crinoidal grain-dominated packstone has been dissolved by late subsurface brines creating a reservoir rock in the Andrews South Devonian Field, West Texas, USA (Lucia 1962).

5.2.4
Effects on Petrophysical Properties Distribution

The effect of shallow diagenesis and simple burial diagenesis on petrophysical properties can usually be related to deposition textures and burial depth or geologic time. Porosity in reefs appears to be substantially reduced by early marine cementation. The Capitan reef of Permian age in the Guadalupe Mountains, West Texas and New Mexico contains up to 70% by volume early radiaxial marine cement suggesting that most of the large construction voids formed with the reef were occluded soon after deposition. Early marine cement apparently occluded large pores in the early reef growth found at the base of Silurian pinnacle reefs in Michigan (Sears and Lucia 1980). Other textures appear to lose porosity gradually through time in a passive margin setting, although in a tectonic setting porosity is lost at a much higher rate.

To illustrate the gradual loss of porosity with time, porosity and fabric data from several limestone studies ranging in geologic time are plotted on a ternary diagram (Fig. 6). The 70% porosity found in modern mud-dominated sediments is reduced after about 10-20 million years of compaction and cementation, but porosity values of between 30 and 40% are still common. After about 150 million years Jurassic grain-dominated limestones retain 25% porosity whereas the porosity of mud-dominated sediments is significantly reduced. In Paleozoic sediments, only grain-dominated fabrics retain sufficient porosity to be considered as reservoir quality, and lower Paleozoic limestones are normally dense.

In this diagenetic environment, the distribution of porosity and permeability after burial would be related to the distribution of depositional textures as presented in Chapter 4. The rate of porosity reduction is greater in mud-dominated textures than in grain-dominated textures. Selective dissolution of aragonite grains will form moldic porosity, cementation of intergrain pore space, and a reduction in permeability (Fig. 7). This dissolution may be in response to the development of local fresh water lenses in association with upward-shoaling depositional sequences, or in response to the dissolution of unstable aragonite and precipitation of intergrain cement in response to burial. The formation of mol-

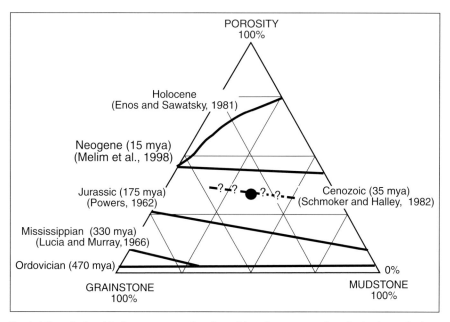

Fig. 6. The relationship between porosity, depositional texture, and geologic time. The loss of porosity with time and burial in a simple burial diagenetic environment is slow. The rate of loss in mud-dominated textures is faster than in grain-dominated textures

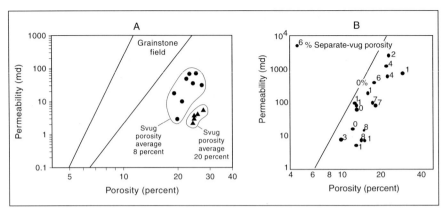

Fig. 7. The effect of separate vugs on permeability in carbonate rocks. **A** Oomoldic porosity in a grainstone. **B** Fossil molds in a dolomitized crinoidal wackestone (Lucia, 1983)

dic porosity in grainstones does not alter the porosity as much as it significantly reduces the permeability

These diagenetic effects will significantly enhance the control of grain-dominated fabrics on reservoir distribution. The permeability of the grain-dominated packstones would be significantly higher than that of the mud-dominated fabrics, and grainstones would have the highest permeability (Fig. 8). The geological age will dictate the porosity and permeability values. Assuming the Jurassic model of the Middle East, the grainstones will average 1000 md, grain-

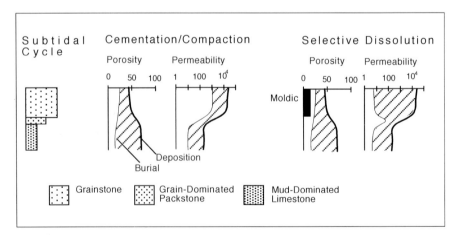

Fig. 8. The reduction in porosity and permeability with burial diagenesis for typical depositional cycles. The porosity and permeability profile of a typical upward-shoaling depositional cycle will be reduced by a time/depth-dependent factor. The burial curves are based on Jurassic Arab D data from Powers (1962). Selective dissolution will not alter the porosity profile but will significantly reduce the grainstone permeability

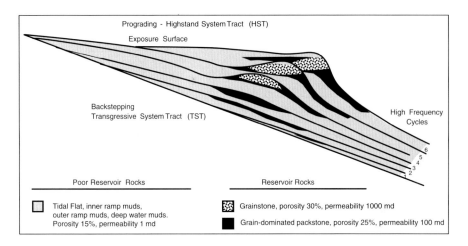

Fig. 9. Distribution of reservoir rocks in high-frequency sequence framework overprinted by cementation, compaction, and selective dissolution

dominated packstones 100 md, and mud-dominated fabrics 1 md. The permeability of moldic grainstones will be between 1 and 10 md.

Permeable grain-dominated fabrics are typically found capping subtidal depositional cycles and are concentrated in the shelf-margin, ramp-crest facies tract (Fig. 9). The close conformance of the diagenetic products to depositional texture allows the use of depositional models for predicting the distribution of petrophysical properties. The principal difference is the very low permeability of moldic grainstones.

5.3
Reservoir Examples

5.3.1
Ghawar (Jurassic) Field, Saudi Arabia

Reservoirs that have a diagenetic history restricted to calcite-cementation, compaction, and selective-dissolution contain a significant volume of the world's hydrocarbon resources. The largest known reservoir, the Ghawar field in Saudi Arabia, belongs to this group. The Ghawar field produces from the upper Jurassic Arab D zone, and although dolostone beds are present, most of the reservoir is limestone. The limestone rock-fabrics range from mudstone to grainstones, and pore types are dominated by intergrain and separate-vug (Powers, 1962). A porosity and permeability cross plot (Fig. 10) of the Arab D in Qatar shows a good conformance between depositional textures, porosity, and permeability (Munn and Jubralla 1987). However, the rock-fabric fields are shifted from the fields described by Lucia (1995) most likely because total porosity rather than interparticle porosity is used. The amount of separate-vug porosity is

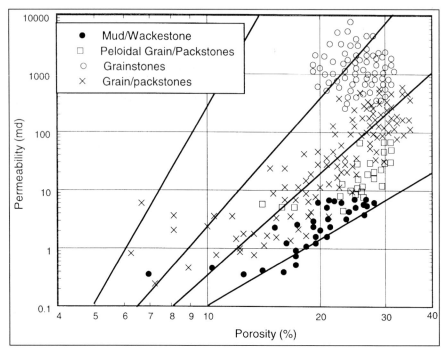

Fig. 10. Porosity-permeability-rock fabric plot showing increase in permeability with increased grain size and sorting in the Arab D, Qatar reservoir of Jurassic age (after Munn and Jubralla 1987). The rock-fabric fields of Lucia (1983) are shown, which are offset because total porosity, not interparticle porosity, is used

not reported. The mud-dominated textures have lost porosity and permeability due to a combination of compaction and cementation (Mitchell et al. 1988). Early cementation of the grainstones inhibited compaction preserving intergrain porosity. Therefore, the spatial distribution of porosity and permeability closely conforms to depositional patterns.

The Arab D zone produces from a high-energy, shallow-water carbonate sequence which is capped by evaporites (Mitchell et al. 1988; Meyer and Price 1993). The sequence is composed of several upward-shoaling cycles. Mitchell et al. (1988) document two cycles whereas Meyer and Price (1993) document six cycles (Fig. 11). No cross sections or maps illustrating the 3-D geometry of the rock fabrics or petrophysical properties are available.

5.3.2
Tubarao (Cretaceous) Field, Offshore Brazil

A recent study of the Cretaceous Albian Tubarao reservoir in the Santos Basin, offshore Brazil by W. Cruz (1997) is another example of a reservoir with a diagenetic history restricted to calcite cementation, compaction, and selective disso-

Fig. 11. Thickness and percent grain- or mud-supported depositional texture distribution for 125 beds in the lower Arab D of Hawiyah-063, depth range 6586-6678 ft. Note the cyclicity in bed thickness and grain-supported textures that divides the interval into six upward shallowing subtidal cycles. (Meyer and Price 1993)

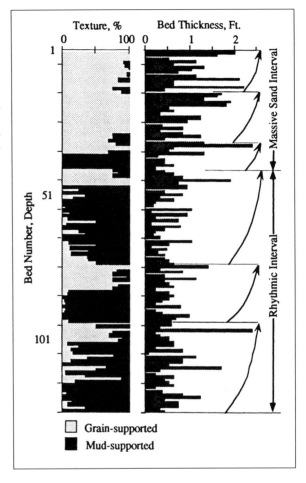

lution. Rock fabrics range from mud-dominated fabrics to grainstone, and pore types from intergrain to separate vug. The principal reservoir fabric is an oncoid grain-dominated packstone/grainstone with an average intergrain porosity of 14% and a permeability of 54 md (Fig. 12a). Mud-dominated packstones and wackestones have an average of 2% porosity and <0.1 md. Ooid grainstones also have an average porosity of 14% but have only an average permeability of 0.4 md because most of the porosity is intragrain microporosity (Fig. 12b).

The diagenetic products can be linked to depositional texture, and patterns of depositional textures can be used to predict the distribution of petrophysical properties. Core description shows a vertical stacking of upward-shallowing subtidal cycles that can be combined into several high-frequency sequences (Fig. 13). The cycles and sequences have been mapped in the field area and form a chronostratigraphic framework for the construction of a petrophysical model (Fig. 14).

Fig. 12. Porosity-permeability-rock fabric plot from the Tubarao field, Santos Basin. **A** The permeability is confined to petrophysical class 2 oncoid grain-dominated packstones (GDP). The GDPs straddle the boundary between class 1 and 2 because the oncoids are very large in size. Mud-dominated fabrics have low permeability values. **B** Ooid grainstone with intragrain microporosity has porosity values similar to oncoid GDP but has less than 1 md permeability. (Data from Cruz 1997)

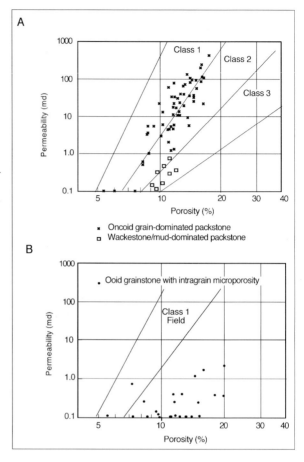

5.3.3
Mississippian Chester Oolite, Oklahoma, USA

The Mississippian Chester oolite of northwestern Oklahoma is an example of a Paleozoic limestone reservoir in which the porosity is controlled by depositional facies and the type of calcite cement (Lucia and Murray 1966). The rock fabrics are composed of four basic constituents: Skeletal fragments, lime mud, calcite cement, and intergrain pore space. The rock fabrics are grainstone, grain-dominated packstone, and mud-dominated fabrics. The reservoir rock is found in some of the grainstones.

The incomplete filling of the intergrain pore space in the grainstones is due to the nature of the calcite cement. Skeletal grains are mostly crinoid fragments. The cement that is commonly found on single-crystal crinoid fragments is a calcite overgrowth called "syntaxial cement" which fills intergrain pore space.

Fig. 13. Vertical stacking patterns from core descriptions of the Cretaceous of the Santos Basin, offshore Brazil (Cruz 1997). Grainstone capped subtidal cycles are grouped into high-frequency sequences based on cycle thickness and the distribution of mud-dominated fabrics. The dominant pore type in the ooid facies is intragrain microporosity that contributes little to permeability. Permeability is concentrated in the oncoid facies because the dominant pore type is intergrain in a grain-dominated packstone. (Cruz 1997)

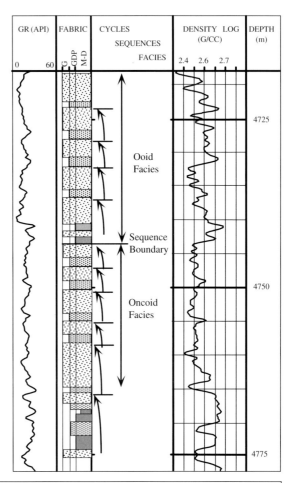

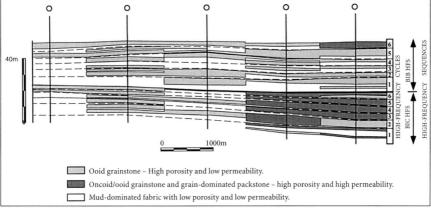

Fig. 14. A reservoir model illustrating the 2-D distribution of reservoir-quality oncoid facies and nonreservoir ooid facies. (Cruz 1997)

Fig. 15. Rock-fabric plot from the Chester Mississippian field showing nonproductive non-oolitic crinoidal grainstones with pore space occluded by syntaxial overgrowth cement, and productive ooid crinoidal grainstones with pore space partially occluded by isopachous cement. Oolitic coatings inhibit the growth of single crystal syntaxial cement and promote the growth of multicrystal isopachous cements. (Data from Lucia and Murray 1966)

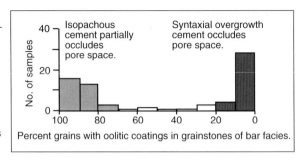

When the crinoids are oolitically coated, an isopachous cement develops instead of the syntaxial cement and this isopachous cement does not fill intergrain pore spaces.

Petrographic studies show that all the productive rocks are grainstones in which 80% or more of the grains are oolitically coated (Fig. 15). Depositional patterns can be used to predict the distribution of petrophysical properties be-

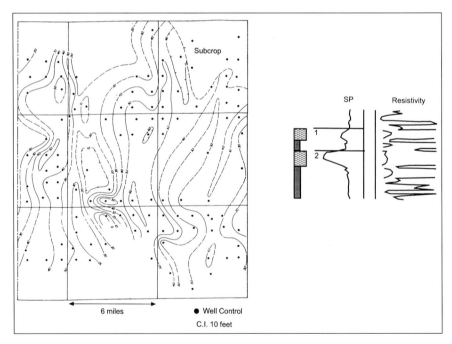

Fig. 16. Isopach map of net bar facies thickness based on the SP curve showing linear trends predicted from modern oolitic grainstone shoal models. Chester Mississippian field, Oklahoma, USA. (Lucia and Murray 1966)

cause of the close conformance between porosity and ooid grainstones. Ooid grainstones are deposited in high-energy environments; commonly in a bar-and-channel pattern. The Mississippian Chester rocks were divided into two facies: the bar facies, which contains ooid grainstone, and the interbar facies, which is characterized primarily by the abundance of less fragmented, non-ooid skeletal debris, the presence of laminated and intermixed lime mud, and numerous shale laminations.

The Spontaneous Potential curve (SP) is used to identify bar and interbar facies from wireline logs. The SP shows high displacement opposite the bar facies and suppressed SP adjacent to the interbar facies. Microlog separation within the bar facies indicates the presence of grainstones with more than 80% ooids. These empirical relationships between facies and log response together with the oolite depositional model were used to map elongated trends of oolite bars that are porous and permeable (Fig.16).

5.3.4
Upper Devonian Reef Buildups, Alberta, Canada

Highly productive Upper Devonian reef buildups are found in the Alberta Basin, Alberta, Canada. These buildups are isolated platforms with several hundreds of feet in relief and up to 150 square miles in area, and are excellent examples of rimmed platforms. In the past, the buildups have been referred to as "reefs". However, they are more correctly called buildups, restricting the term "reef" to bindstone facies. The margins have relatively steep slopes into the basin and contain bindstone reef facies characteristic of rimmed platforms. Many have been completely dolomitized. The Redwater and Swanhills buildups are limestone, and production is controlled by the distribution of depositional facies.

The Redwater buildup is the largest, covering about 150 square miles (Fig. 17A). Production is located on the structurally high northeastern margin on the buildup. Klovan (1964) presents a detailed description of this area. The key facies is the organic reef (bindstone: Fig. 17) which is composed of stromatoporoids that overgrow each other to form a bound fabric. Tabulate corals are also present and brachiopods are abundant. The reef is located at the shelf margin. A stromatoporoid detritus facies (rudstone) is also found at the shelf margin (Fig. 17A) and is interpreted to be a reef broken up by strong wave action. Fore-reef deposits are wackestones with scattered stromatoporoid and coral fragments. The back reef is composed of seven facies. A peritidal facies (laminite), composed of irregular laminations and fenestral fabric is located on the paleotopographic high of the buildup, and an *Amphipora* facies composed of wackestones and grain-dominated packstones is located in the interior of the buildup.

The Swanhills buildup is in an area of numerous Upper Devonian buildups. The facies progression is similar to Redwater but the buildup is smaller, covering about 40 square miles (Fig. 18). A stromatoporoid bindstone/rudstone with a matrix of coarse-grained skeletal grainstones and packstones is located at the

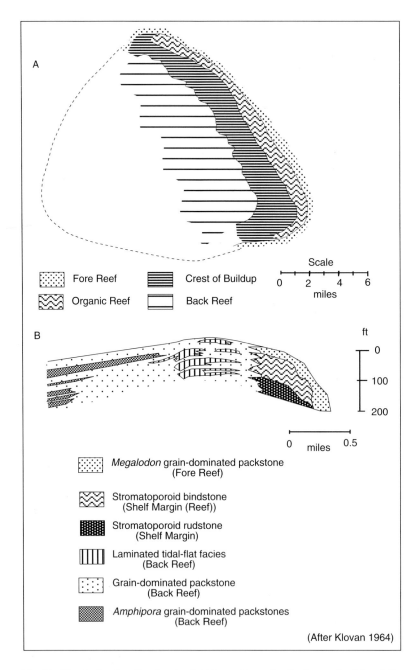

Fig. 17. The facies distribution in the Upper Devonian Redwater buildup (Klovan 1964).
A A map showing buildup outline and general depositional facies patterns. **B** A cross section through the northeastern margin of the buildup showing general facies patterns

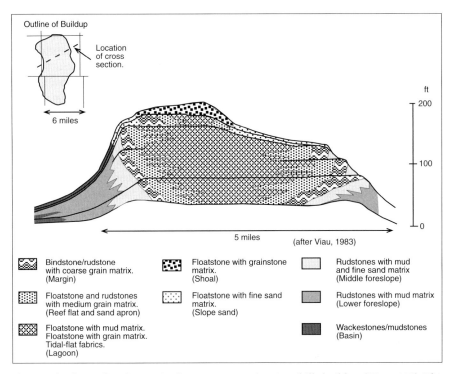

Outline of Buildup

Location of cross section.

6 miles

5 miles

(after Viau, 1983)

ft

200

100

0

Bindstone/rudstone
with coarse grain matrix.
(Margin)

Floatstone and rudstones
with medium grain matrix.
(Reef flat and sand apron)

Floatstone with mud matrix.
Floatstone with grain matrix.
Tidal-flat fabrics.
(Lagoon)

Floatstone with grainstone
matrix.
(Shoal)

Floatstone with fine sand
matrix.
(Slope sand)

Rudstones with mud
and fine sand matrix
(Middle foreslope)

Rudstones with mud matrix
(Lower foreslope)

Wackestones/mudstones
(Basin)

Fig. 18. The facies distribution in the Upper Devonian Swanhills buildup (Viau 1983). The most common large fossils are stromatoporoids and *Amphipora*

margin of the buildup, and reef debris mixed with skeletal grains and mud are found in the fore slope facies (Fig. 18). The volume of mud increases downslope, and the basin facies is composed of wackestones and mudstones. Immediately behind the reef are stromatoporoid floatstones and rudstones with a matrix of grainstone and packstone. The interior of the buildup is composed of *Amphipora* floatstones with a matrix varying from mud-dominated to grain-dominated fabrics (Fig. 18). The buildup is divided into six growth stages, and the interior of the upper stage is composed of stromatoporoid floatstone with a grainstone matrix and flanking beds of floatstone with a fine sand matrix.

Intergrain and intragrain pore types are the most common in both of these buildups. Moldic porosity is present in the form of dissolved aragonite grains, and fenestral porosity is present in back-reef tidal flat facies. No massive dissolution fabrics have been described although exposure surfaces are present, and only minor dolomite is reported. Therefore, compaction, cementation, and selective dissolution are the dominant diagenetic processes, and the distribution of petrophysical properties should conform to patterns of depositional textures.

Geological descriptions of rimmed platforms are dominated by detailed descriptions of fossil material because the organisms are the key indicators of depositional environment. The facies patterns and depositional environments in these two buildups are described on the basis of buildup topography and various types of stromatoporoids, *Amphipora*, and corals. Typically, little attention has been paid to the grain size that controls the pore size. The fossils occur as large particles, giving rise to the terms "rudstone" and "floatstone". However, the description of the sediment between the large fossil fragments is of principal interest when characterizing the petrophysical properties for reservoir characterization.

The descriptions of both buildups suggest the presence of grain-dominated fabrics in the vicinity of the bindstone reef and mud-dominated fabrics in the interior and fore slope facies. The facies are typically composed of large fossil fragments with a matrix of grains and mud, and the particle size, sorting, and interparticle porosity of the matrix controls the pore-size distribution. In the vicinity of the bindstone (reef) facies, the volume between large particles is normally filled with sand-sized particles and the sand size controls the pore size. Occasionally the volume between large particles is pore space and the pore size is controlled by the size of the large particles. In the interior (lagoon) and down the flanks, the volume between the large particles is normally filled with mud, but beds are occasionally present where the matrix is grain-dominated packstone.

The highest production rates are typically from wells that penetrate the margin of the buildup where intergrain porosity in the grain-dominated matrix of rudstones is concentrated. Production from the interior is from discontinuous beds of grain-dominated fabrics.

5.3.5
Moldic Grainstone, Permian, Guadalupe Mountains, USA

The results of a study on a body of moldic grainstone is included here as an example of the effect of selective dissolution on the distribution of petrophysical properties. In Chapter 4 the Algerita Escarpment outcrop study was used to illustrate the concentration of grainstones in the progradation, HST. Although the section is completely dolomitized, the fabrics closely mimic depositional textures. Data collected from the outcrop show that the grainstones have average permeability values of 100 md, grain-dominated packstones 10 md, and wackestones 1 md, whereas all three fabrics have porosity values close to 13%.

The exception to this is the grainstone in cycle 7, where selective grain dissolution has inverted the pore structure, resulting in areas of high porosity (average 20%) but low permeability (average 2.5 md) within the cross-bedded grainstone (Hovorka et al. 1993; Fig. 19). The formation of the moldic porosity is interpreted as resulting from a local fresh-water lens that formed in response to either the overlying tidal flat deposits or the overlying high-frequency sequence boundary. Therefore, the pattern of the moldic grainstone is not only linked to the depositional pattern of the grainstone, but also to local paleogroundwater movements.

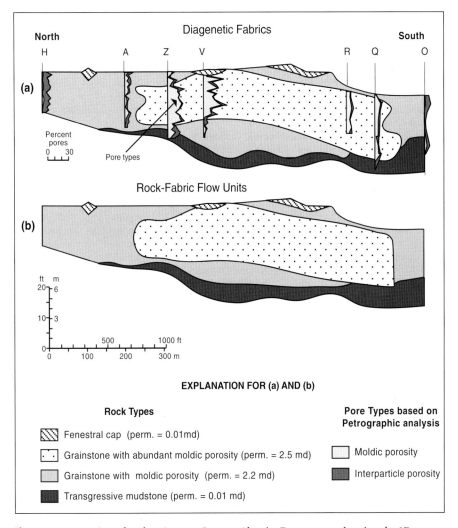

Fig. 19. Cross section of cycle 7, Lawyer Canyon, Algerita Escarpment showing the 2D geometry of a moldic grainstone body. Selective dissolution of grains in the grainstone has resulting in significantly lower permeability than would be expected. (After Kerans et al. 1994)

References

Anselmetti FS, Eberli GP (1993) Controls on sonic velocity in carbonates. Pageoph 141, 2/3/4: 287–323

Budd DA, Land LS (1990) Geochemical imprint of meteoric diagenesis in Holocene ooid sand, Schooner Cays, Bahamas: correlation of calcite cement geochemistry with extant groundwaters. J Sediment Petrol 60, 3: 361–378

Budd DA, Hammes U, Vacher HL (1993) Calcite cementation in the upper Floridian aquifer: a modern example for confined-aquifer cementation models?, Geology 21, 1: 33–37

Cruz WM (1997) Study of Albian carbonate analogs: Cedar Park Quarry, Texas, USA, and Santos Basin reservoir, southeast offshore Brazil. Unpubl PhD thesis, The University of Texas
at Austin, Austin, Texas

Dix GR, Mullins HT (1988) Rapid burial diagenesis of deep-water carbonates: Exuma Sound,
Bahamas. Geology 16, 8: 680–683

Enos P, Sawatsky LJ (1981) Pore networks in Holocene carbonate sediments. J. Sediment Petrol 51, 3: 961–985

Goldhammer RK (1997) Compaction and decompaction algorithms for sedimentary carbonates. J Sediment Res 67, 1: 26–56

Harris PM, Kendall CGSTC, Lerche I (1985) Carbonate cementation: a brief review. In: Schneidermann JS, Harris PM (eds) Carbonate cements. SEPM Spec Publ 36: 79–95

Heydari E, Moore CH (1989) Burial diagenesis and thermochemical sulfate reduction, Smackover Formation, southeastern Mississippi salt basin. Geology 17, 12: 1080–1084

Heydari E, Moore CH, (1993) Zonation and geochemical patterns of burial calcite cements:
Upper Smackover Formation, Clarke County, Mississippi. J Sediment Petrol 63, 1: 44–60

Hovorka SD, Nance HS, and Kerans C (1993) Parasequence geometry as a control on porosity
evolution: examples from the San Andres and Grayburg Formation in the Guadalupe
Mountains, New Mexico. In: Loucks RG, Sarg JF (eds) Carbonate sequence stratigraphy:
recent developments and applications. AAPG Mem 57: 493–514

Hurley NF, Lohmann KC (1989) Diagenesis of Devonian reefal carbonates in the Oscar Range,
Canning Basin, Western Australia. J Sediment Petrol 59, 1: 127–146

James NP, Ginsburg RN (1979) The seaward marginal Belize barrier and atoll reefs. Int Assoc
Sedimentol Spec Publ 3. Blackwell Scientific Publication, Oxford London Edinburgh Bos-
ton Melbourne, 191 pp

Kerans C., Lucia FJ, Senger RK (1994) Integrated characterization of carbonate ramp reservoirs using outcrop analogs. AAPG Bull 78, 2: 181–216

Klovan FE (1964) Facies analysis of the Redwater Reef Complex, Alberta, Canada. Bull Can Pet
Geol 12, 1: 1–100

Lucia FJ (1962) Diagenesis of a crinoidal sediment: J Sediment Petrol 32, 4: 848–865

Lucia FJ (1983) Petrophysical parameters estimated from visual description of carbonate
rocks: a field classification of carbonate pore space. J Pet Technol March: 626–637

Lucia FJ (1995) Rock/fabric petrophysical classification of carbonate pore space for reservoir
characterization. AAPG Bull 79, 9: 270–300

Lucia FJ, Murray RC (1966) Origin and distribution of porosity in crinoidal rocks. Proc 7th
World Petroleum Congress, Mexico City, Mexico, 1966, pp 409–423

Melim LA, Anselmitte FS, Eberli GP (1998) The importance of pore type on permeability of
Neogene carbonates, Great Bahama Bank. (in press)

Meyer FO, Price RC (1993) A new Arab-D depositional model, Ghawar field, Saudi Arabia. SPE
Middle East Oil Techn Conf, Bahrain, SPE 25576, pp 495–474

Mitchell FJ, Lehman PJ, Contrell DL, Al-Jallal IA, Al-Thagafy MAR (1988) Lithofacies, diagenesis,
and depositional sequence; Arab-D member, Ghawar field, Saudi Arabia. In: Lomando AJ,
Harris PM (eds) Giant oil and gas fields: a core workshop. SEPM core workshop 12,: 459–514

Moore CH, Druckman Y (1981) Burial diagenesis and porosity evolution, upper Jurassic
Smackover, Arkansas and Louisiana. AAPG Bull 65, 4: 597–628

Munn D, Jubralla AF (1987) Reservoir geological modeling of the Arab D reservoir in the Bul
Hanine Field, offshore Qatar: approach and results. SPE Middle East oil Show, Bahrain,
March 7-10, 1987, SPE 15699, pp 109–120

Powers RW (1962) Arabian Upper Jurassic carbonate reservoir rocks. In: Ham WE (ed) Classification of Carbonate Rocks: a symposium. AAPG Mem 1: 122–192

Schmoker JW, Halley RB (1982) Carbonate porosity versus depth: a predictable relation for
south Florida. AAPG Bull 66, 12: 2561–2570

Schmoker JW, Krystinic KB, Halley RB (1985) Selected characteristics of limestone and dolomite reservoirs in the United States. AAPG Bull 69, 5: 733–741

Sears SO, Lucia FJ (1980) Dolomitization of northern Michigan Niagaran reefs by brine reflux-
    ion and freshwater/seawater mixing. In: Zenger DH, Dunham JB, Ethington RL (eds) Con-
    cepts and models of dolomitization. SEPM Spec Publ 28: 215–236
Shinn EA (1969) Submarine lithification of Holocene carbonate sediments in the Persian Gulf.
    Sedimentology 12: 109–144
Swirydezuk K (1988) Mineralogical control on porosity type in upper Jurassic Smackover
    ooid grainstones, southern Arkansas and northern Louisiana. J Sediment Petrol 58, 2: 339–
    347
Viau C (1983) Depositional sequences, facies and evolution of the Upper Devonian Swan Hills
    reef buildup, Central Alberta, Canada. In: Harris PM (ed) Carbonate buildups – A core
    workshop. SEPM Core Workshop 4: 112–143
Weyl PK (1959) Pressure solution and the force of crystallization – a phenomenological theo-
    ry. J Geophys Res 63, 11: 2001–2025

Diagenetic Overprinting and Rock-Fabric Distribution: The Dolomitization/Evaporite-Mineralization Environment

6.1
Introduction

In Chapter 5 we stated the basic premise that the three-dimensional spatial distribution of petrophysical properties is controlled by the spatial distribution of geological processes, processes that can be separated into depositional and diagenetic. In Chapter 4 we discussed depositional processes and the relationship between the spatial distribution of depositional textures and petrophysical properties. Reservoir studies have made it abundantly clear that the petrophysical properties found in carbonate reservoirs are significantly different from those of Holocene carbonate sediments due to various diagenetic processes.

A key issue in predicting and mapping the distribution of petrophysical properties within a carbonate reservoir is how the diagenetic overprint modifies depositional textures and impacts the distribution of petrophysical properties. The premise here is that the degree of diagenetic conformance to depositional textures is linked to geochemical, hydrological issues. If the transport of material in and out of the system is not a major factor in producing the diagenetic product, the product will generally conform to depositional patterns. Nonconformance exists if the transport of ions in and out of the system by fluid flow is required to produce the diagenetic product. In this case, knowledge of the hydrological system may be required to map the diagenetic products. Because we are interested in geometric patterns, key geochemical, hydrological information is the location of the fluid source, the direction of fluid flow, and the geochemical changes that occur along the flow path.

Basic diagenetic processes are here grouped according to their conformance to depositional patterns. As discussed in Chapter 5, the products of cementation, compaction, and selective dissolution are grouped because they can normally be linked to depositional textures. Compaction and associated cementation is a diagenetic process that is a function of rock strength and the time-overburden history. Selective dissolution of unstable aragonite allochems and associated precipitation of calcite cement does not require the transport of material by fluid flow, although the local introduction of meteoric water may enhance the process. Early cementation processes require fluid flow to import calcium and carbonate into the system, but the fluid flow is marine water flowing through permeable grainstones and reefs by tidal and wave energy. Late or burial cementation may be formed by chemical compaction and thus be related to a depositional environment, or it may require regional transport of ions by ground water, in which case it need not be related to a depositional environment.

In this chapter, we will focus on reflux dolomitization and evaporite mineralization. Dolomitization requires fluid flow for the introduction of magnesium into the system. Therefore fluid flow is an important factor in the origin of dolomite fabrics, and the dolomite pattern may not conform to depositional patterns. Dolomitization may increase the particle size, significantly modifying the pore-size distribution of the sediment and smoothing out important petrophysical differences in depositional textures. Predolomite diagenetic history may significantly alter the permeability structure and result in dolomitizing waters following diagenetic rather than depositional flow paths.

Diagenetic gypsum and anhydrite are commonly associated with dolomitization and require the transport of sulfate into the system by high sulfate, hypersaline water. Studies have shown little linkage between depositional facies patterns and patterns of diagenetic gypsum or anhydrite. However, sulfate commonly selectively occludes pore space in highly permeable grainstones, whereas it typically forms poikilotopic anhydrite in other fabrics.

In Chapter 7 we will discuss massive dissolution, collapse brecciation, and fracturing. The products of this diagenetic environment typically exhibit little relationship to depositional textures, being controlled primarily by precursor diagenetic events and ground water flow. Nonfabric selective, massive dissolution clearly requires the reorganization of pore space through the removal of carbonate from some areas and the precipitation of carbonate in other areas through a complicated geochemical, hydrological meteoric flow system. Massive dissolution processes significantly modify rock fabrics and create touching-vug pore systems that have little relationship to depositional patterns.

6.2
Dolomitization/Evaporite Mineralization

The processes of hypersaline dolomitization and evaporite mineralization comprise a diagenetic environment characterized by refluxing hypersaline evaporated seawater from the surface down into underlying strata replacing sea water and interacting with groundwater (Fig. 1). Hypersaline tidal flat environments and associated hypersaline ponds, lakes, and lagoons are common sources for hypersaline water. The high density of the saline water and the elevation of the tidal flats produces the hydrodynamic potential needed for saline water to flow down and seaward through the underlying sediment, producing dolomitization and the precipitation of gypsum and anhydrite.

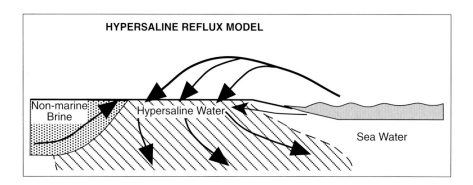

Fig. 1. Hypersaline reflux model based on modern data from the Trucial Coast and Qatar, Middle East. Seawater transported onto an arid tidal flat evaporates and flows downward through the underlying tidal flat and subtidal sediments converting the calcium carbonate sediment to dolomite. Hypersaline seawater mixes with the regional ground water landward, forming a mixed seawater/meteoric water hypersaline fluid

6.2.1
Dolomitization

The most important aspect of dolomitization to affect reservoir performance is the increase in pore size resulting from the increase in particle size from fine crystalline to medium or coarse crystalline in mud-dominated fabrics.

Dolomitization is a diagenetic process that converts limestones to dolostones through a microchemical process of calcium carbonate dissolution and dolomite precipitation. Two equations for dolomite formation are shown below. They form end members of a continuous series of possible reactions.

$$2CaCO_3 + Mg^{++} = CaMg(CO_3)^2 + Ca^{++} \text{ (replacement)} ,$$

$$Mg^{++} + Ca^{++} + 2CO_3 = CaMg(CO_3)^2 \text{ (cementation)}.$$

From the replacement equation, it can be seen that the Mg/Ca ratio in the water is a controlling factor in the reaction. Analysis of waters associated with both ancient and modern dolomite indicates that a Mg/Ca ratio of higher than 1 is required for dolomitization of limestones (Folk and Land 1975) (Fig. 2). Normal marine water, however, has a Mg/Ca ratio of 7 without forming significant quantities of dolomite, demonstrating the impact of kinetics on dolomitization. Marine dolomite is commonly found associated with evaporated marine water with Mg/Ca ratios of between 10 and 50. These increased ratios result from loss of calcium through gypsum or anhydrite precipitation in response to evaporation of sea water (Fig. 3).

In a closed system, the replacement of calcite or aragonite by dolomite will reduce mineral volume because dolomite has a smaller molar volume than either calcite or aragonite. However, dolomitization involves the flow of large quanti-

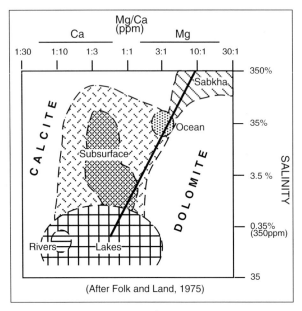

Fig. 2. Relationship of dolomite and calcite stability fields to Mg/Ca ratios and salinity (Folk and Land 1975). Dolomitizing water is found in a variety of environments including Sabkha (tidal flat), groundwater, and saline lakes

Fig. 3. The increase in Mg/Ca molar ratio during evaporation of sea water due to the precipitation of $CaSO_4$ (gypsum or anhydrite). Data is from Bonaire, N.A., and the curve is computed from gypsum and aragonite precipitation curves. (Deffeyes et al. 1965)

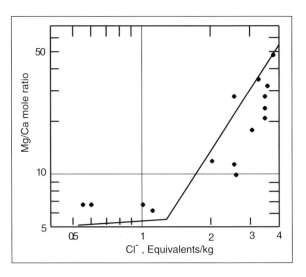

ties of water through the rock, which requires an open system. Thus, not only magnesium but also carbonate can be added to the system, resulting in loss of porosity by dolomite cementation. Therefore, the dolomitization process in-

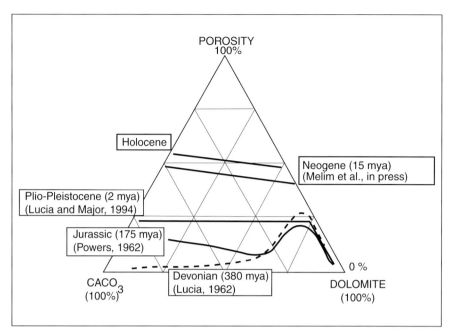

Fig. 4. Relationship between porosity, dolomitization, and time. Limestone is more porous in young carbonates than dolomites and less porous in older carbonates. The loss of porosity in limestones relative to dolomites with time and burial is probably due to differential compaction

cludes both replacement of calcite or aragonite by dolomite and cementation of pore space by dolomite cement. This process is known as overdolomitization.

Modern dolomitic sediments show little change in pore volume with increased amounts of dolomite (Fig. 4). Plio-Pleistocene dolostones, however, are commonly less porous than their limestone counterparts. A dolomitization study of Plio-Pleistocene carbonate on the island of Bonaire in the Netherlands Antilles demonstrated that the limestone has an average porosity of 25 % whereas the dolomite has an average porosity of 11 %, suggesting a loss of porosity with dolomitization by overdolomitization (Lucia and Major 1994). If dolomitization occurred in a closed system and only magnesium was substituted for calcium in the carbonate lattice, the resulting dolomite would have an average porosity of 35 % (Fig. 5). In contrast, Paleozoic dolostones are usually more porous than their associated limestones. Comparing limestone and dolomite porosity values by geological age shows a gradual change from limestone, being more porous than dolomite, to dolomite being more porous than limestone (Fig. 4). This relationship is also seen in the depth plot of porosity from South Florida (Fig. 6; Schmoker and Halley 1982). Dolomite is less porous than limestone at shallow depths and in Tertiary age strata. Dolomite porosity does not decrease as rapidly as limestone porosity with depth, and this results in dolomite having higher po-

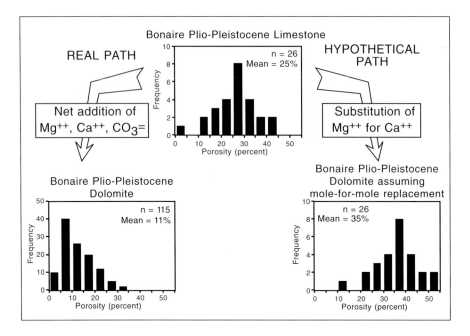

Fig. 5. Porosity frequency plots for dolomite, the precursor limestone, and for a hypothetical dolomite calculated from the limestone frequency plot assuming mole-for-mole replacement. (Lucia and Major 1994)

rosity than limestone at deeper Cretaceous strata. This reversal may be due to differential compaction between limestone and dolostone and to preferential calcite cementation of limestones over dolomites.

Dolomitization can change the rock fabric significantly (Fig. 7). In mud-dominated fabrics the dolomite crystals may be the same size as the mud-sized particles, or the dolomite crystals may be much larger than the mud-sized particles they replace. Dolomite crystals commonly range in size from several to 200 μm, whereas the crystals in calcium carbonate mud are usually less than 20 μm in size. Thus, dolomitization of a carbonate mud can result in an increase in crystal size from less than 20 μm to 200 μm with a corresponding increase in pore size. The increase in particle size caused by dolomitization of mud-dominated fabrics will greatly increase the flow characteristics of the rock and improve the capillary properties because of the corresponding increase in pore size. The pore size increase resulting from a larger particle size is offset by the addition of dolomite resulting from the importation of magnesium and carbonate with dolomitizing waters. The net addition of material results in an overall loss of porosity and a corresponding decrease in pore size.

The increase in pore size results from the reorganization of pore space during dolomitization. For example, 200-μm dolomite crystals will occupy space once occupied not only by the less than 20-μm calcite or aragonite crystals, but also

Fig.6. Relationship of porosity to depth and lithology in South Florida. The shallowest data are from Quaternary carbonates and the deepest data are from the Cretaceous. Dolomite becomes more porous than limestone at a depth of 6000 ft, which approximates the Tertiary-Cretaceous boundary. (Schmoker and Halley 1982)

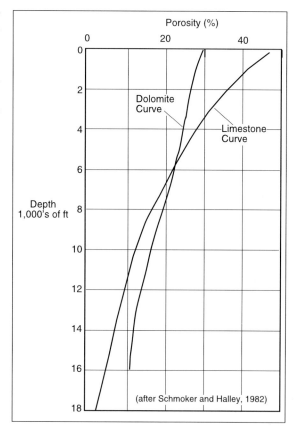

by the micro pore space between those crystals. Thus, the 200-μm dolomite crystals are both replacing the calcite crystals and occluding the micro pore space between the crystals. Because the dolomite crystal occupies more space than preexisting calcite crystals, calcium, magnesium, and carbonate must be transported to the site of the growing dolomite crystal. There are two basic sources for the calcium, magnesium, and carbonate ions; 1) dissolution of nearby carbonate crystals and 2) regional ground water. These sources are referred to as local and distant sources by Murray (1960). The dissolution of nearby carbonate produces intercrystal porosity, whereas precipitation of dolomite from regional ground water produces dolomite cement (Fig. 7).

Vuggy porosity is created similarly during dolomitization. Once dolomite crystals form, they become preferred sites for dolomite precipitation. Therefore, carbonate dissolved from one location will be transported to an existing growing dolomite crystal. Skeletal fragments are usually the last to be dissolved during dolomitization because of their size, and when the carbonate is transported to an existing growing dolomite crystal, a dissolved skeletal fragment is left as a fossil mold, a type of separate vug (Fig. 7). Particle dissolution and transportation of ions from particles to intercrystal pore space may have little effect on to-

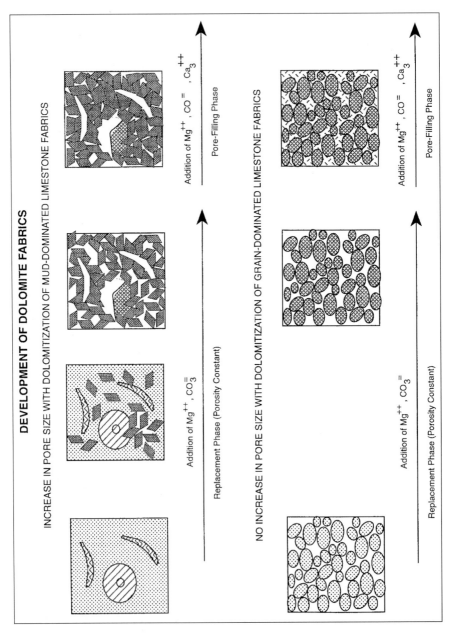

Fig. 7. The change in pore structure during dolomitization of **A** mud-dominated limestone and **B** grain-dominated limestone. The conversion of mud-dominated limestone to medium or large crystalline dolomite increases pore size, whereas the dolomitization of a grain-dominated fabric has little effect on pore size

tal porosity but the reduction in intercrystal porosity and associated reduction in pore size will reduce permeability because moldic pore space does not contribute significantly to permeability.

Grainstones are usually composed of grains much larger than the dolomite crystal size (Fig. 7), so that dolomitization does not have a significant effect on their pore-size characteristics. Grain-dominated packstones are also typically composed of large grains, but replacement of the lime-mud fraction with larger than 100-mm dolomite crystals will have a great impact on their petrophysical properties.

In summary, dolomitization affects the flow characteristics of carbonate reservoir rocks by (1) increasing particle size, (2) decreasing pore volume due to a net addition of dolomite, (3) developing moldic pores, and (4) increasing resistance to compaction.

6.2.1.1
Calcitization of Dolomite

The dolomite replacement process is reversible, resulting in calcification of dolomite in the presence of high calcium waters. Ground water with a Mg/Ca ratio less than 1 is capable of dissolving dolomite and reprecipitating calcite. Models for the formation of low Mg/Ca ratio waters capable of calcitizing dolomite are highly varied, but all involve meteoric or deep ground waters. The dissolution of $CaSO_4$ by meteoric ground water creates a low Mg/Ca water that is capable of calcitization (Lucia 1972; Back et al. 1983). Calcium-rich water that originates from deep saline deposits and migrates upward along faults is also capable of calcitizing dolomite (Land and Prezbindowski 1981).

Calcitized dolomite crystals commonly display calcite centers and dolomite rims. Hollow dolomite rhombs (often refered to as skeletal dolomite) have been observed in many dolostone bodies, suggesting that centers of dolomite crystals are more soluble than dolomite rims. Dissolution of dolomite crystals normally results in a separate-vug pore type. In some studies petrographic observations suggest that the calcite is occluding hollow dolomite rhombs, and in other studies the calcite appears to be replacing the dolomite (Evamy 1967).

6.2.2
Evaporite Mineralization

Anhydrite ($CaSO_4$) and gypsum ($CaSO_4 \cdot 2H_2O$) are evaporite minerals commonly found in dolomite reservoirs. Halite (NaCl) is uncommon but may be found occluding pore space adjacent to salt beds. Gypsum is the common evaporite mineral found in modern sediments and at shallow depths. The change from gypsum to anhydrite is controlled by temperature and the activity of water (Fig. 8). Recent anhydrite is found in Persian Gulf tidal-flat sediments, where the surface temperatures are very high and the interstitial waters very saline. The increase in temperature with depth results in near-surface gypsum converting to

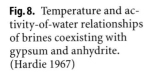

Fig. 8. Temperature and activity-of-water relationships of brines coexisting with gypsum and anhydrite. (Hardie 1967)

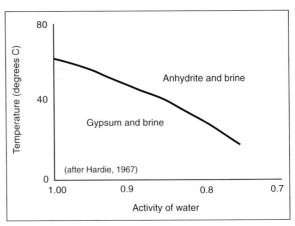

anhydrite. In the Permian Basin of West Texas, gypsum is found down to depths of 4000 ft.

Four types of anhydrite are commonly found in dolomite reservoirs (Fig. 9). Poikilotopic anhydrite is found as large crystals of anhydrite with inclusions of dolomite and is often distributed randomly throughout the rock. Crystals form by a combination of replacement and pore-filling mechanisms, and porosity is reduced in proportion to the amount of pore-filling. Poikilotopic anhydrite is typically scattered and unevenly distributed. Therefore, between the crystals the matrix maintains its original pore-size distribution. Permeability and capillary properties are related more to pore size than to porosity and are only slightly modified by the formation of poikilotopic anhydrite.

Nodular anhydrite is found in dolostone in the form of microcrystalline masses of anhydrite, and commonly forms within the sediment by displacement as either anhydrite or gypsum. Therefore, it is a diagenetic texture and should not be used to interpret characteristics of the depositional environment. The nodules commonly make up a small percentage of the bulk volume and have little effect on porosity or permeability.

Pore filling anhydrite is typically pervasive and reduces both porosity and pore-size distribution of carbonate reservoir rocks because it occludes intergrain, intercrystal, and vuggy pore space.

Bedded anhydrite is found in laterally continuous beds a few inches to hundreds of feet thick, and is deposited out of a hypersaline body of water as gypsum and later converted to anhydrite. It can be either laminated or composed of coalesced nodules. Coalesced nodular anhydrite may form (1) by precipitation out of a body of water as gypsum or (2) by displacement and replacement of near-surface sediment as either gypsum or anhydrite. Bedded anhydrite commonly acts as a reservoir barrier or seal.

Halite is also found as a pore-filling evaporite mineral in carbonate reservoirs associated with bedded halite. Halite is very soluble and easily dissolved out of subsurface samples during sample preparation. It is isotropic and clear under petrographic examination, making it difficult to identify in thin section. Impregnating the sample with blue plastic is very helpful in distinguishing halite from pore space.

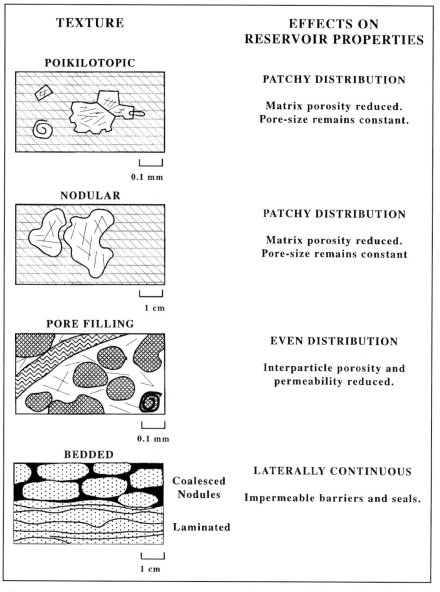

Fig. 9. Diagram showing basic anhydrite and gypsum textures found in carbonate rocks and their effect on reservoir properties

6.2.2.1
Calcitization of Anhydrite/Gypsum

Anhydrite or gypsum may be replaced by calcite and occasionally by native sulfur and authigenic silica (Lucia, 1972). Outcrops of Permian age ramp and shelf strata in the Guadalupe Mountains, Texas and New Mexico, contain calcite pseudo-morphs of nodular, and poikilotopic anhydrite, suggesting that anhydrite commonly found in subsurface equivalents has been replaced by calcite during uplift in the presence of shallow meteoric waters (Lucia 1961; Scholle et al. 1992). Unusually light carbonate isotopic values found in some calcites suggest replacement by organic carbon, probably associated with hydrocarbons. Native sulfur is occasionally found, associated with calcite replacement of anhydrite nodules in the subsurface equavalents, suggesting the presence of sulfate-reducing bacteria.

6.2.3
Effects on Petrophysical Properties Distribution

The pattern of dolomitization will be controlled by the volume and flow path of the dolomitizing water. The source of the hypersaline water is the supratidal environment and the volume of dolomitizing water will be related to the volume of hypersaline water produced. If a small volume of hypersaline water is formed, only the tidal flat facies and a small volume of subtidal facies will be dolomitized. If large volumes of dolomitizing water are produced, hundreds of meters of underlying subtidal sediment can be dolomitized (Fig. 10).

     The flow path will be controlled by permeability. The permeability effect is observed at the boundaries of dolomite bodies where finger flow into more permeable strata produce strataform bodies of dolomite. The permeability effect is also observed in late dolomites where permeability is restricted to one rock fabric or facies. Examples of this are selective dolomitization of mud-dominated

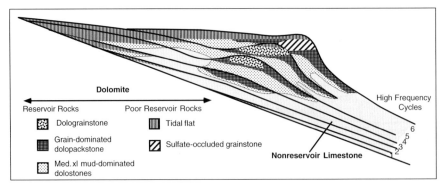

Fig. 10. Generalized distribution of reservoir rocks in a high-frequency sequence resulting from cementation-compaction followed by hypersaline reflux dolomitization and evaporite mineralization. The increase in particle size in mud-dominated dolostone increases the potentially productive volume

facies over tightly cemented grainstone facies, and selective dolomitization of fractured and brecciated karsted facies and fault zones.

Where hypersaline dolomitization has occurred, mud-dominated fabrics can be reservoir rocks as well as grain-dominated fabrics increasing the reservoir potential. Whereas mud-dominated limestone fabrics are typically nonreservoir rocks, mud-dominated dolostones can be highly porous and permeable because of their resistance to compaction and larger particle and pore sizes.

The petrophysical class may not conform to the depositional texture because the class of dolostone is determined by depositional texture (modified Dunham categories) and dolomite crystal size. Two factors that affect dolomite crystal size are (1) the saturation with respect to dolomite as reflected by Mg/Ca ratios and (2) the surface area of the precursor fabric.

In the evaporative reflux model, hypersaline water supersaturated with dolomite enters the system from the tidal flat surface and flows down and out to the sea. The saturation with dolomite will decrease along the down-and-out flow path, and the dolomite crystal size should increase along the flow path. Modern dolomites found in evaporite tidal flats have a crystal size of around 5 μm, as do dolomitized tidal-flat sediments in the geological record. In the San Andres and Grayburg dolomite reservoirs of West Texas and New Mexico, the dolomite crystal size can often be correlated with distance below peritidal facies. In the Permian age Seminole San Andres reservoir, West Texas, dolomite crystal size increases in size with depth (Fig. 11). The reservoir is capped by a peritidal facies

Fig. 11. Cross plot of depth versus dolomite crystal size for mud-dominated and grain-dominated dolostones in the Seminole San Andres reservoir, West Texas showing overall increase in crystal size with depth

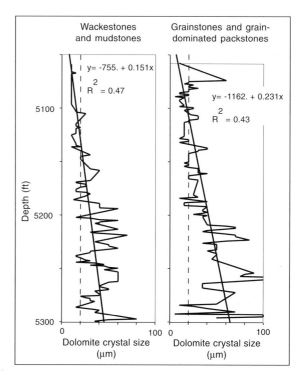

tract, and the increase in dolomite crystal size is interpreted reflecting a decrease in dolomite saturation as the dolomitizing water flows from the peritidal facies down.

Dolomite crystal size is also correlated with precursor texture. Dolomitized grains are commonly coarser crystals than dolomitized mud. The most extreme case is pseudomorphic replacement of echinoderm fragments, producing millimeter-size dolomite crystals in a matrix of 200 mm crystals (Lucia 1962). Surface area can be correlated with the amount of carbonate mud in the rock and perhaps is a key parameter in control of dolomite crystal size. The effect of grain content on dolomite crystal size can be seen the Seminole depth plot (Fig. 11) in that the grain-dominated fabrics have larger dolomite crystals than the mud-dominated fabrics even though the average crystal size increases with depth. The effect of surface area may also partially explain the observation that late dolomites are always medium to large size. Predolomite burial cementation and compaction reduces porosity and surface area, establishing the conditions required for the formation of larger dolomite crystals.

Dolograinstone, grain-dominated dolopackstone with crystal size less than 100 μm, and mud-dominated dolostone with crystal size less than 20 μm have petrophysical properties similar to the limestone precursor. Dolostones with less than 20-μm dolomite crystals are concentrated in tidal flat and adjacent facies. Within these strata, the petrophysical classes will conform to the depositional textures, and depositional patterns can be used to map the spatial distribution of petrophysical properties (Fig. 12). The dolomite crystal size of mud-dominated dolostones located farther from tidal flat facies are typically larger

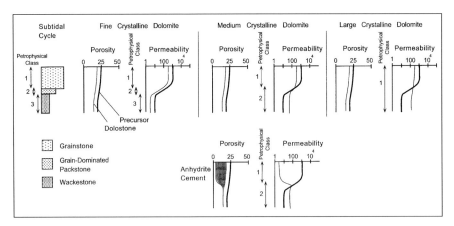

Fig. 12. Illustration of the effects of hypersaline dolomitization on porosity and permeability profiles in a single subtidal upward-shallowing cycle. The precursor limestone curves are based on Jurassic Arab D data from Powers (1962). The porosity of the precursor will be reduced during the dolomitization process by an amount related to the volume throughput of dolomitizing water. The permeability profile will change dependent upon (1) the change in porosity and (2) the resulting dolomite crystal size in the mud-dominated fabrics. The vertical stacking of petrophysical classes will change with the change in dolomite crystal size

than 20 µm, perhaps larger than 100 µm. Within these strata, the petrophysical class of the mud-dominated dolostones will not be the same as the precursor mud-dominated limestone and the petrophysical classes will not conform to depositional textures (Fig. 12). If the dolomite crystal size is less than 100 µm, grainstone depositional patterns can be used to map the spatial distribution of petrophysical properties. If the dolomite crystal size is greater than 100 µm, depositional patterns may not be useful.

Evaporitic tidal-flat environments are a source of sulfate as well as magnesium-rich waters. Poikilotopic and nodular forms have little effect on reservoir properties. Bedded evaporites are commonly found in tidal flat facies tracts and form reservoir seals and barriers. Pore filling anhydrite is commonly found in grainstone bodies associated with thick tidal flat facies, grainstone of the prograding, HST, and can be mapped according to depositional patterns.

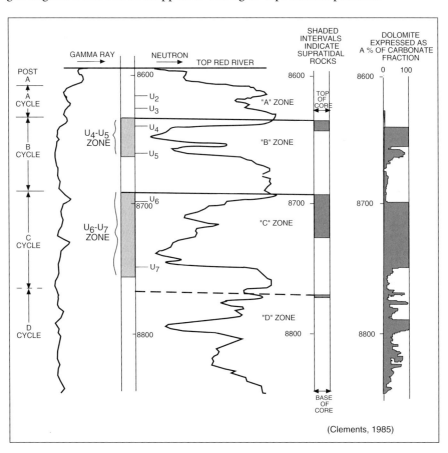

(Clements, 1985)

Fig. 13. Gamma-ray/neutron log showing depositional cycles and the principal upper Red River production zones related to dolomite and supratidal rocks. Limestone is dense and dolomite porous, and the dolomite conforms to the supratidal rocks and the underlying subtidal. Red River (upper Ordovician) Pennel field, Montana. (Clement 1985)

6.3
Reservoir Examples

6.3.1
Red River Reservoirs, Montana and North Dakota

The Red River reservoirs are Upper Ordovician in age, and are examples where limestone is tight and dolostones are the reservoir rocks. Within the dolostones, there is good conformance to depositional texture, but the limestone-dolomite boundary (the dolomite replacement front) is difficult to predict because it requires knowledge of the precursor permeability structure.

Red River reservoirs are located in subtidal and supratidal sediments deposited on a very broad shallow carbonate shelf (Clement 1985). There are three tidal-flat capped cycles in the Red River formation (Fig.13). The supratidal facies in the lower sequence (C Zone) is a thick anhydritic dolostone with beds of bedded and nodular anhydrite. Dolomite extends varying distances beneath this facies into the subtidal facies, probably depending on the flux of hypersaline water formed by evaporation of sea water in the supratidal environment (Fig. 14). The subtidal dolostone is more coarsely crystalline than the supratidal dolostone and commonly more porous. Thus, the thickness of productive dolostone is a function of the thickness of subtidal dolostone. Rapid lateral changes from dolostone to limestone are common and difficult to predict.

The middle sequence (B Zone) is a main producing horizon on the Cedar Creek Anticline. The supratidal facies is thin and is capped by bedded and coalesced nodular anhydrite. The supratidal is dolomitized and dolomitization extends varying distances into the underlying subtidal sediments. The sequence is divided into three productive dolomite facies; the supratidal, intertidal and subtidal. These facies can be recognized on porosity and resistivity logs because of their different particle and pore-size characteristics. The supratidal and intertidal facies tend to be laterally continuous whereas the subtidal dolomite zone is discontinuous.

6.3.2
Andrews South Devonian Field, West Texas

The Andrews South Devonian field is an example of a late dolomite which conforms well with depositional textures. It is located in Andrews County, West Texas (Lucia 1962). It produces from carbonates deposited near the shelf margin during Devonian times. The sequence is upward-shoaling from mud-dominated to grain-dominated and is overlain by a regional unconformity. The productive facies are porous limestones and dolostones. The sequence is divided into a lower pellet packstone unit and an upper crinoidal grainstone unit. The units were calibrated with the gamma ray log. The lower unit has an irregular pattern of porosity due to the lack of current activity in the depositional process. The upper grainstone unit is divided into three facies.

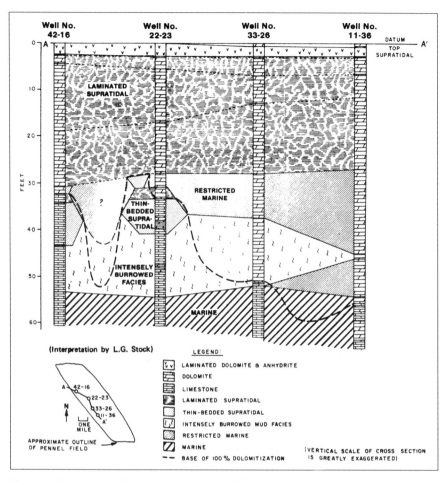

Fig. 14. Relationship of lithology to depositional facies U6-U7 zone, Red River Formation. The dolomite-limestone contact does not conform with subtidal facies because it is a function of the flow paths of dolomitization water and not patterns of sedimentation. (Clement 1985)

The grainstone facies is dominated by crinoid fragments and cemented tight with syntaxial calcite cement. The grain-dominated packstone facies is a grain-supported sediment with some intergrain lime mud and is porous and permeable due to dissolution of the intergrain lime mud. The dolomite facies is a coarsely crystalline, dolomitized wackestone and is the most permeable facies in the reservoir. The facies are calibrated with porosity (Fig. 15). The dolomitizing water was probably formed in an overlying supratidal environment and flowed down into these sediments. The crinoidal grainstones were cemented by that time and impermeable. The wackestones were still porous and permeable because they were not buried very deep. Therefore, the dolomitizing water prefer-

Fig. 15. Generalized relationship among percent of lime mud, porosity, limestone, and dolomite. In the Andrews South Devonian field, West Texas, porosity logs can be used to map rock-fabric facies. (Lucia 1962)

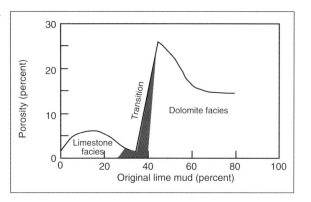

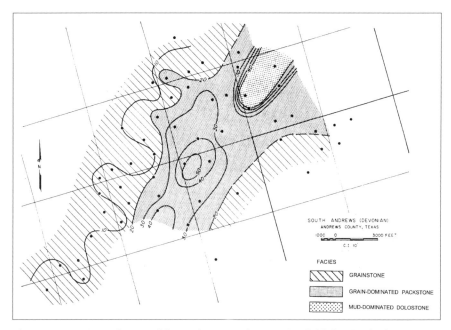

Fig. 16. Net pay isopach map of the Andrews South Devonian field showing highest net pay in the dolomitized mud-dominated facies, some net pay in the limestone grain-dominated packstone (inner bar) facies, and little net pay in the grainstone (bar) facies. (Lucia and Murray 1966)

entially flowed through the wackestones, converting them to dolomite. The fabric selective nature of the dolomitizing process allows depositional models to be used directly to map reservoir characteristics and predict other occurrences (Fig. 16). Rapid lateral changes from low or nonproductive limestones to highly productive dolomites can be expected.

6.3.3
Flanagan (Upper Clear Fork) Field, Gaines County, West Texas

The Clear Fork, San Andres, and Grayburg Permian reservoirs of West Texas and New Mexico are characterized by thick intervals of hypersaline reflux dolomite, intervals that are hundreds to thousands of feet thick. Permeable zones are scattered throughout the thick dolomite intervals. In some intervals the dolomite fabrics conform well with depositional textures whereas in others the textural differences are overridden by dolomitization and sulfate emplacement.

The conformance between depositional textures and rock fabrics in the Flanagan field is moderate. There are two tidal-flat capped cycles in the reservoir (Lucia, 1972; Fig. 17), called here the lower cycle and the upper cycle. The reservoir is concentrated in the subtidal facies of the upper cycle whereas the subtidal facies of the lower cycle is dense. Also, only the central portion of the upper subtidal facies is productive, and the upper subtidal facies loses porosity westward forming the western edge of the reservoir.

There are two petrophysical rock fabrics within the subtidal; medium crystalline dolowackestone and medium crystalline grain-dominated dolopackstone. The distribution of grain-dominated packstone was mapped using core control and compared with productivity (Fig. 18). Whereas the net pay isopach (porosity-feet) within the upper cycle presented little variability, an isopach of grain-dominated packstone presented a definite pattern, which matched the highest recovery per well in the reservoir.

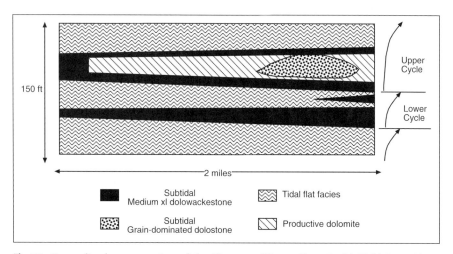

Fig. 17. Generalized cross section of the Flanagan (Upper Clear Fork) Field, West Texas, showing an upper and lower tidal-flat capped cycle. The productive zones are found in the subtidal portion of the cycles, and to that degree are conformable to sedimentary patterns. In detail, however, the conformance is less predictable because portions of the subtidal have been overdolomitized, probably due to concentrated fluid flow, and are dense. (After Lucia 1972)

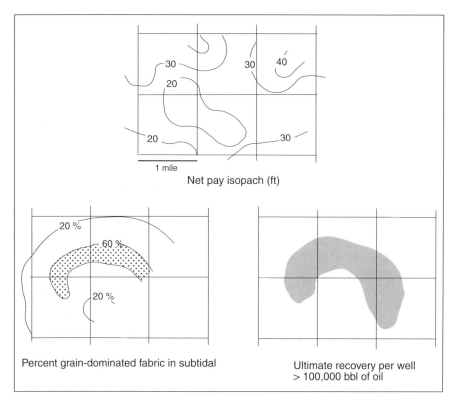

Fig. 18. Isopach maps of the Flanagan Upper Clear Fork field, West Texas, showing that well productivity is linked to the distribution of the grain-dominated facies in the upper cycle (see Fig. 17). In this case, productivity is linked to sedimentation patterns

Therefore, there is a degree of conformance between diagenesis and sedimentation in that the tidal flat is a tight, finely crystalline laminated dolostone, and the reservoir is a subtidal medium crystalline dolostone. However, there is only moderate conformance between depositional textures and rock fabrics in the subtidal facies in that, the wackestones are productive in some areas and over dolomitized in others.

References

Back W, Hanshaw BB, Plummer LN, Rahn PH, Rightmire CT, Rubin M (1983) Process and rate of dedolomitization: mass transfer and 14C dating in a regional carbonate aquifer. GSA Bull 94, 12: 1415–1429

Clement JH (1985) Depositional sequences and characteristics of Ordovician Red River Reservoirs, Pennel Field, Williston Basin, Montana. In: Roehl RO, Choquette PW (eds) Carbonate petroleum reservoirs. Springer, Berlin Heidelberg New York, pp 85–106

Deffeyes KS, Lucia FJ, Weyl PK (1965) Dolomitization of Recent and Plio-Pleistocene sediments by marine evaporite waters on Bonaire, Netherlands Antilles. In: Pray LC, Murray RC (eds) Dolomitization and limestone diagenesis – a symposium. SEPM Spec Publ 13: 71–88

Evamy BD (1967) Dedolomitization and the development of rhombohedral pores in limestones. J Sediment Pet 37: 1204–1215

Folk RL, Land LS (1975) Mg/Ca ratio and salinity: two controls over crystallization of dolomite. AAPG Bull 59, 1: 60–68

Hardie LA (1967) The gypsum-anhydrite equilibrium at one atmosphere pressure. Am Mineral 52: 171–200

Land LS, Prezbindowski DR (1981) The origin and evolution of saline formation water, lower Cretaceous carbonates, South-Central Texas, USA. J Hydrol 54: 51–74

Lucia FJ (1961) Dedolomitization in the Tansill (Permian) Formation. Geol Soc Am Bull 72: 1107–1110

Lucia FJ (1962) Diagenesis of a crinoidal sediment. J Sediment Petrol 32, 4: 848–865

Lucia FJ (1972) Recognition of evaporite-carbonate shoreline sedimentation. In: Rigby JK, Hamblin WK (eds) Recognition of ancient sedimentary environments. SEPM Spec Publ 16: 160–191

Lucia FJ, Major RP (1994) Porosity evolution through hypersaline reflux dolomitization. In: Purser B, Tucker M, Zenger D (eds) Dolomites: a volume in honour of Dolomieu. Int Assoc Sedimentol Spec Publ 21: 309–324

Lucia FJ, Murray RC (1966) Origin and distribution of porosity in crinoidal rock. Proc the World Petroleum Congr, Mexico City, Mexico, 1966, pp 406–423

Melim LA, Anselmitte FS, Eberli GP (1998) The importance of pore type on permeability of Neogene carbonates, Great Bahama Bank. (In press)

Murray RC (1960) Origin of porosity in carbonate rocks. J Sediment Petrol 30: 59–84

Powers RW (1962) Arabian Upper Jurassic carbonate reservoir rocks. In: Ham WE (ed) Classification of carbonate rocks, AAPG Mem 1: 122–192.

Schmoker, JW, Halley RB 1982 Carbonate porosity versus depth: a predictable relation for south Florida. AAPG Bull 66, 12: 2561–2570.

Scholle PA, Ulmer DS, Melim LA (1992) Late stage calcites in the Permian Capitan Formation and its equivalents, Delaware Basin margin, West Texas and New Mexico: evidence for replacement of precursor evaporites. Sedimentology 39: 207–234

Diagenetic Overprinting and Rock-Fabric Distribution: The Massive Dissolution, Collapse, and Fracturing Environment

7.1
Introduction

A key question in mapping diagenetic effects is the degree of conformance between diagenetic products and depositional patterns. As discussed in Chapter 5, the products of cementation, compaction, and selective dissolution can normally be linked to depositional textures because the transport of material in and out of the system is not an important factor in producing the diagenetic product. However, if the transport of ions in and out of the system by fluid flow is required to produce the diagenetic product, then the product may not conform to depositional patterns. In this case, knowledge of the geochemical, hydrological system may be required to map the diagenetic products, including the source of the fluid and the direction of fluid flow.

In Chapter 6, we discussed reflux dolomitization and evaporite mineralization, two processes that require fluid flow for the introduction of magnesium and sulfate into the system. The source of the diagenetic fluid is located in the tidal flats environment and hypersaline bodies of water; the direction of flow is downward and seaward from these locations; and the chemical changes that occur along the flow path include loss of dolomite and sulfate saturation. An increase in dolomite crystal size can be found along the flow path, significantly modifying the pore-size distribution of the sediment and smoothing out important petrophysical differences found in depositional textures. Predolomite diagenetic history may significantly alter the flow paths and result in dolomitizing waters following diagenetic rather than depositional fabrics.

In this chapter we will discuss massive dissolution, collapse brecciation, and fracturing. Massive dissolution refers to non-fabric selective dissolution, including enlargement of fractures, dissolution of bedded evaporite minerals, and cavern formation at any scale. Most commonly this process is thought to be related to the flow of near-surface ground water, here referred to as the meteoric diagenetic environment, but often included under the general heading of karsting. However, some massive dissolution is thought to be associated with deep fluid movements and unrelated to surface events (Dravis and Muir 1993). Deep dissolution is suggested to be related to "thermochemical sulfate reduction", a process of producing sulfuric acid by the reduction of sulfate minerals or the H_2S in hydrocarbons (Heydari and Moore 1989; Kaufman et al. 1990). In this chapter, we will only deal with the meteoric environment, although the reservoir properties resulting from massive dissolution in both the meteoric and "burial" environment are likely to be similar.

Collapse brecciation and fracturing is included in this diagenetic environment because it is associated with the collapse of large caverns. Fractures are a common feature of carbonate rocks and are formed in response to stress which can be generated in many ways. Three common stress regimes are tectonic, geopressure, and the formation of large voids (caves). Tectonic fractures form in response to the bending and breaking of strata, and fracture distribution can be related to bending moments and distance from faults. In some instances, fracture density is related to lithology, for instance dolomite tends to be more fractured than limestone. Geopressure fractures are formed when pore fluids are trapped and cannot be expelled as overburden pressure increases. As the pore pressure approaches overburden pressure, fracture pressure is reached and the rock breaks. Fractures form in the roofs of large caves as burial increases the overburden pressure to a point where the roof fails and collapse breccia forms. Fracture distribution is directly related to cave distribution and indirectly related to tectonic fractures and the water table. In this chapter we will only consider fractures formed by massive dissolution and subsequent collapse.

The products of this diagenetic environment are controlled by precursor diagenetic events, tectonic fracturing, and ground water flow. Nonfabric selective, massive dissolution clearly results in the reorganization of pore space through the removal of carbonate from some areas and the precipitation of carbonate in other areas through a complicated geochemical, hydrological meteoric flow system. Rock fabrics are significantly modified and touching-vug pore systems are created which have little relationship to depositional patterns. Massive dissolution creates a pore system that cannot be related to interparticle or separate-vug porosity, and is referred to as a touching-vug pore system.

Massive dissolution reservoirs are commonly composed of touching vugs and a matrix composed of interparticle and separate-vug pore space, the bimodal pore system. Matrix pore space may be related to depositional fabrics if it has not been altered by later diagenetic events. Touching-vug systems typically control fluid flow because of their wide conduits but typically contribute less than 1% to reservoir porosity. Most of the reservoir pore space is located in matrix pore types.

Although massive dissolution, collapse, and fracturing commonly occur long after deposition, they are not the last diagenetic event. Diagenesis begins after sedimentation and never ends. The diagenetic events that follow dissolution and collapse, and which require fluid flow, will be spatially controlled by the touching-vug pore system. A common product of this system is late stage dolomite. The origin and source of late stage dolomitizing fluids is speculative. However, outcrop studies have shown that the touching-vug pore system created by massive dissolution and collapse controls the flow paths of the dolomitizing fluids. Thus, many late stage dolostones do not conform to depositional patterns, but they do conform to breccia and fractures formed by massive dissolution and collapse.

7.2
Massive Dissolution, Collapse, and Fracturing

Massive dissolution is a characteristic process associated with major fresh water aquifers, the meteoric diagenetic environment (Fig. 1). Meteoric water falling on a land surface will enter the groundwater system and flow toward the sea. Between the earth's surface and the water table, the pore space is partially saturated with capillary held water except during heavy rains or floods. This zone of partial water saturation is called the *vadose zone*. Flow in this zone occurs during rain fall and flooding, is focused at fracture intersections and dolines, and is dominantly in a down direction. Below the water table, the pore space is 100% saturated with water and is called the *phreatic zone*. Flow in this zone can have both a lateral and a downward component and is focused along fractures and in dissolved passageways.

The location of the boundary between the vadose and phreatic zone is not constant but moves up and down based on the balance between rain fall recharge, evapotranspiration, and discharge into bodies of water. The thickness of the vadose zone is dependent upon topography. In mountainous topography the vadose zone can extend to a depth of over 1000 ft, whereas in flat country it is typically a few feet to 200 ft thick. Most reservoirs are found in the passive margin tectonic environment where exposure surfaces have only a small amount of relief. Therefore, we will focus our remarks on meteoric diagenetic environments associated with surfaces that have little topography.

Climate will have an effect on the interaction between ground water and sea water as groundwater flow enters the coastal regime. In humid climates, ground water that reaches the coastline and encounters sea water rides on top of the interstitial brine, forming a fresh water wedge because the fresh water is less dense than sea water. The effect may be the formation of coastal cave systems due to the mixing of waters with different degrees of $CaCO_3$ saturation. In arid climates,

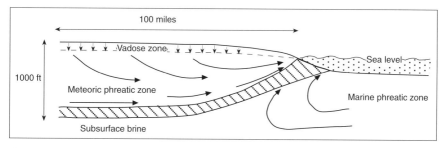

Fig. 1. The meteoric-groundwater environment showing groundwater zones, mixing zone, and fluid flow directions. Dissolution is known to be concentrated in the vadose zone, the upper levels of the meteoric phreatic zone, and in the mixing zone

however, the groundwater flow will encounter a zone of refluxing hypersaline marine water near the shoreline and form a zone of mixed hypersaline and meteoric water. Dolomitization and sulfate mineralization will occur instead of massive dissolution.

Dissolution and precipitation occur throughout this environment. Pendulous and meniscus cements form in capillary trapped water in the vadose zone and sparry cements form in the vadose zone. Because rain water is undersaturated, it initially dissolves carbonates in the vadose and upper phreatic zones. Studies of the Florida Aquifer have discovered undersaturated phreatic water hundreds of feet below the water table (Back 1963). This suggests that dissolution is not restricted to the vadose and upper phreatic zones, but continues at a slower rate throughout the phreatic zone.

Geochemical studies of have shown that the mixing of two waters saturated with calcite at differing CO_2 partial pressures can produce solutions that are undersaturated with calcite, and thus gain a capacity to dissolve calcite (Plummer 1975). Sea water is supersaturated with $CaCO_3$, but when mixed with a more dilute saturated solution, such as meteoric groundwater, it may become undersaturated (Fig. 2; Plummer 1975). Studies of the Florida Aquifer and coastal caves in Yucatan, Mexico have demonstrated limestone dissolution at the mixing zone between sea water or subsurface brine and fresh water of the aquifer (Stoessell et al. 1989).

Undersaturated meteoric water is characteristic of the meteoric diagenetic environment, and dissolution is focused in the vadose and high phreatic zones. Studies of karsting and modern caves have focused on the vadose zone. Karsted landscapes are characterized by internal drainage into dolines or sink holes formed by concentrated dissolution, perhaps at the intersection of fracture sys-

Fig. 2. Saturation index of calcite in mixtures of solutions saturated with calcite at differing CO_2 partial pressures and 25 °C and modified seawater. (Plummer 1975)

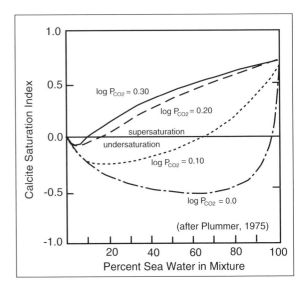

tems. Dolines lead to underground passages that may contain underground streams and are commonly floored by cave sediment and collapse breccia. These caves are concentrated in the vadose and upper phreatic zones. Cave maps commonly show that fracture systems exert a strong control on cave geometry. Whereas there may be considerable porosity in the limestone, the fluid flow is concentrated in the high permeability fracture and cavern systems. Studies of the Edwards Aquifer in Texas have shown that fracture and cave permeability is 100 times higher than matrix permeability. Little is known about the extent of modern caves below the water table, but geological studies suggest that they may extend hundreds of feet below the water table.

Preexisting tectonic fracture patterns and the position of the vadose/phreatic contact clearly have strong controls on dissolution patterns in the meteoric diagenetic environment. Depositional textures have little control on the distribution of massive dissolution, however beds of soluble minerals, such as the evaporite minerals anhydrite, gypsum, or halite, can be selectively dissolved in the ground water environment, forming cave systems. Also, limestone beds may be preferentially dissolved over dolomite beds in strata where these two lithologies are interbedded.

Massive dissolution and karsting can be directly associated with subaerial exposure surfaces because these surfaces are the point of entry of the water into the ground water system. In the reflux dolomitization model, identifying supratidal surfaces and evaporite lagoons as the point of entry of the dolomitizing water is important for predicting dolomite patterns and dolomite porosity. Similarly, identifying the point of entry of the meteoric groundwater system is important for predicting the distribution of massive dissolution products such as caves, collapse breccias, and fractures.

A review of sequence stratigraphy studies suggests that karsting is not well developed at the boundaries of high-frequency cycles. Small karst pits filled with overlying sediment may be found at exposure surfaces bounding high-frequency sequences and composite sequences (Kerans et al. 1994). Massive dissolution effects, however, appear to be associated with second order unconformities, exposure surfaces with millions of years missing (Budd et al. 1995).

Identifying the surface to which the formation of collapse breccia and caverns is linked can be difficult because massive dissolution can form hundreds of feet below a subaerially exposed surface. There is no typical vertical succession of karst fabrics that can be used to identify the karst surface, and one-dimensional data, such as vertical cores, may have undisturbed strata between the karsted surface and the first underlying collapse breccia. The tendency for observers to place the surface of exposure immediately above collapse breccias has led to a misinterpretation that major dissolution occurs at cycle boundaries. Outcrop observations, however, do not confirm this conclusion.

The formation of caves and cavern systems by massive dissolution initiates a series of events that have an important effect on reservoir geometry (Fig. 3). The first is the importation of sediment into the caves by flowing surface water. This sediment can be deposited in the vadose caves or filter down through the frac-

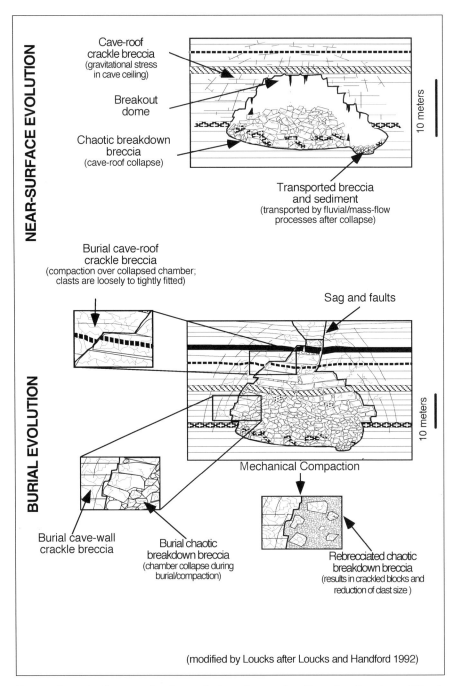

Fig. 3. Schematic diagrams showing evolution of cave-related breccias. (Loucks and Handford 1992)

tures and passages in the phreatic zone. Large caves cannot support themselves for long periods of time and are destroyed by roof collapse. Roof collapse can occur during the time of exposure and can continue after exposure as overlying sediment is deposited and the overburden pressure increases. The history of collapse is often very complicated and extends over hundreds of millions of years (Loucks and Handford 1992).

Products of collapse important for reservoir description are fractures and collapse breccia; the formation of fracture porosity and interbreccia-block pore space referred to as touching-vug pore types. The geometry of collapse breccia will follow the pattern of major cave development. Fracture porosity will be concentrated in the failed roof and flanks of the cavern system (Kerans 1988; Loucks and Handford 1992).

7.2.1
Effects on Petrophysical Properties Distribution

The distribution of petrophysical properties in the massive dissolution, collapse, and fracture environment is controlled by preexisting fracture patterns, meteoric flow regime, and stratum lithology (Fig. 4). Massive dissolution creates large, connected voids herein referred to as touching vugs. If the vugs are large enough to form caverns and caves, they can collapse, fracturing the overly roof and forming a collapse breccia with interbreccia-block porosity. Cavern development and collapse can produce a vertical sequence of fracture and breccia types. Fracture breccia, mosaic breccia, or crackle mosaic are characteristic of the collapsed roof. Chaotic breccia with open interclast pore space or filled with internal sediment are characteristic of the collapsed cavern. Open and sediment-

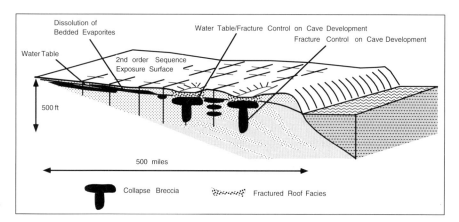

Fig. 4. Patterns of touching-vug porosity formed by massive dissolution, cavern development, and collapse. Geometry of touching-vug bodies is controlled by strataform evaporite bodies, water table, and precursor fracture patterns

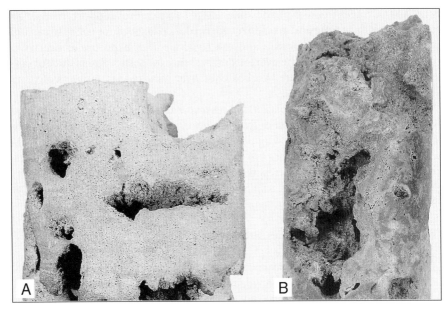

Fig. 5. Examples of cavernous porosity observed in cores. **A** Miami Oolite from the Miami aquifer, Miami, Florida. **B** Niagaran (Silurian), Michigan pinnacle reef

filled vugs and small caverns are characteristic of strata below and adjacent to the collapsed caverns.

Massive dissolution may initially provide a huge increase in porosity on a very local scale, but the large caverns produced rarely persist far into the geological record. Cave infill and cave collapse normally destroys much of the cavernous porosity. Massive-dissolution reservoirs often have lower porosity than other carbonate reservoirs. The pore sizes, however, are typically very large and transmit fluid easily (Fig. 5). Thus, massive dissolution processes often result in excellent permeability and high oil saturation despite the low porosity.

7.3
Reservoir Examples

The massive dissolution diagenetic environment commonly is a late diagenetic event and follows cementation, compaction, selective dissolution, reflux dolomitization, and sulfate mineralization. Therefore, massive dissolution reservoirs often retain some of the rock fabrics from these earlier diagenetic events. Two San Andres fields from West Texas are presented as examples of the combination of depositionally-controlled petrophysical distribution and massive-dissolution-controlled brecciation, collapse, and fracturing. A Mississippian reservoir from Wyoming is presented as an example of a petrophysical distribution controlled by collapse-fracture-controlled late dolomitization. Although we focus on hydrocarbon reservoirs, some of the most interesting "reservoir" examples are lead and zinc mines found in the Ordovician of Missouri, USA and in the Devonian of western Canada.

7.3.1
San Andres Fields, West Texas

Although many San Andres fields in West Texas produce from interparticle and separate-vug pore systems, some of the major fields have an overprint of caverns and fractures that have a major effect on fluid flow. Two of the reservoirs are the giant Yates Field with 4 billion barrels on oil originally in place, and the interesting Taylor Link field with 50 million barrels in place. These two fields are only about 15 miles apart and have had a similar depositional and diagenetic history. The Yates field is divided into a lower package of low-energy open-ramp facies and an upper package composed of three-four upward shallowing high-frequency sequences (Tinker and Mruk 1995). The upper three HFSs are composed of 13 upward shallowing high-frequency cycles. The top of the San Andres is a significant karst surface.

Although the Yates field is dolomite, there is good conformance between depositional textures and reservoir quality in the Yates field with grainstones and grain-dominated packstones having the highest matrix porosity (Tinker and Mruk 1995). The depositional textures have a predictable relationship to the Permian paleogeography of the region, with grainstones concentrated at the ramp crest and mud-dominated fabrics and peritidal sediments concentrated in the middle and inner ramp.

An extensive fracture network overprints the matrix porosity. The fractures important to reservoir performance are believed to have formed early in the life of the reservoir and act as conduits for solution enlargement during post San Andres karstification. Fracture patterns are predictable because they have a well defined orthogonal NW-SE and NE-SW orientation (Fig. 6). Numerous caves have been mapped using bit drops and caliper logs, and these caves are found hundreds of feet below the San Andres top (Fig. 7). There is some indication that cave formation occurred in response to freshwater lenses that developed during exposure of each of the four HFSs. The Yates field is relatively shallow and most caves have not undergone mechanical collapse.

The Taylor Link field is similar to the Yates field in that the matrix petrophysical properties conform well to depositional textures and the fractures do not (Lucia et al. 1992). The field is composed of an upper dolograinstone unit that is about 60 ft thick and probably represents an amalgamation of several upward-shoaling cycles. The pore space is between ooid grains, and the porosity permeability transform for this facies is within the class 1 field (Lucia 1995). The lower unit is a fusulinid fine crystalline dolowackestone with scattered fusumolds. Fractures and large circular vugs are concentrated in the dolowackestone. Fracture types include simple fractures, solution-enlarged fractures, and micro breccia and associated fractures.

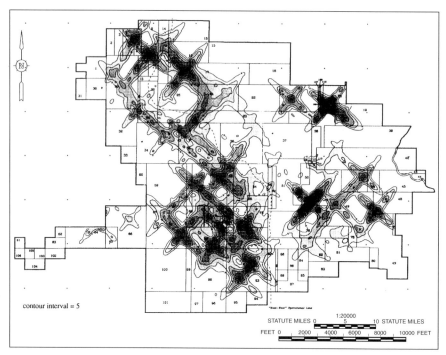

Fig. 6. Isopach of the number of fractured feet in a 20 ft slice map of the Yates San Andres field showing NW-SE and orthogonal NE-SW trends imposed on data (Tinker and Mruk 1995)

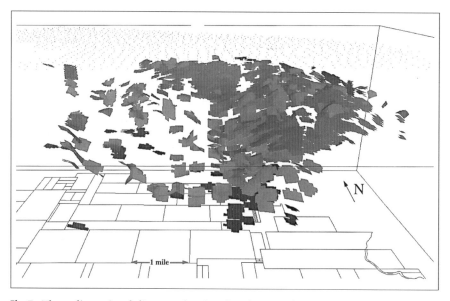

Fig. 7. Three-dimensional diagram showing distribution of caves in the Yates San Andres field. (Tinker and Mruk 1995)

Fig. 8. Depth plot from a well in the Taylor Link San Andres field showing matrix and fracture permeability. Fracture permeability is shown in black as the difference between matrix permeability estimated from fabric analysis and total permeability from core measurements. (Lucia et al. 1992)

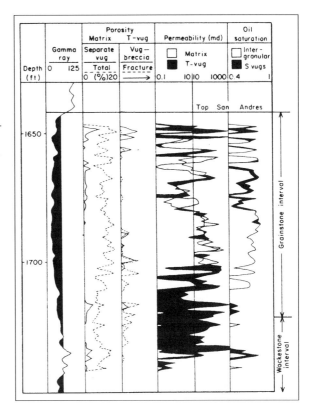

Core analyses show highest porosity values in the dolograinstone facies whereas permeability values are as high in the dolowackestone as in the dolograinstone facies (Fig. 8). This is due to fracture porosity in the wackestone. An estimate of the fracture permeability was obtained by estimating the matrix contribution to total permeability using the porosity-permeability transform for a grainstone and comparing the values with core values, values that include both matrix and fracture permeability. The results show that all the permeability in the grainstone interval (about 20 md) can be attributed to intergrain porosity, whereas all the permeability in the wackestone (about 20 md) can be attributed to fracture porosity. Fracture porosity is estimated at 1%.

A reservoir model of the Taylor Link reservoir (Fig. 9) shows the upper unit composed of a porous and permeable grainstone facies which grades laterally into low porosity fractured wackestone. The lower unit is composed entirely of low porosity fractured wackestone. It has little porosity or oil saturation, but has fracture permeability and is a thief zone.

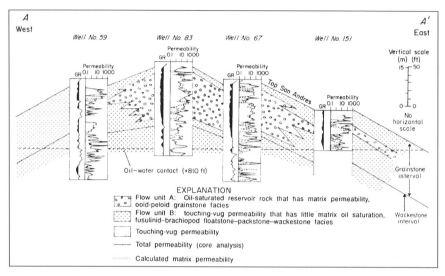

Fig. 9. Northwest-southeast cross section of the Taylor-Link San Andres reservoir showing distribution of ooid grainstone rock-fabric facies and fractured fusulinid wackestone facies. (Lucia et al. 1992)

7.3.2
Elk Basin Field, Wyoming-Montana

The Elk Basin field produces from the Mississippian Madison Group and is located in the north end of the Big Horn basin on the border between Wyoming and Montana. The reservoir rocks were deposited on a shallow carbonate shelf in an upward-shoaling sequence from mud-dominated to grain-dominated sediments. Evaporite beds capped several of these cycles. At the end of Madison deposition, the Madison was exposed subaerially and an extensive karst surface formed during Late Mississippian and Early Pennsylvanian time. The evaporite beds were dissolved to form strataform breccias and caverns, and vertical caverns were formed probably along fracture planes (Fig. 10). During deposition of the overlying Pennsylvanian sediments, the cavities were partially filled with a mixture of clay and sand. Collapse of the caverns resulted from burial. As a result of infill and collapse, large areas of the Elk Basin field are isolated from each other.

An excellent outcrop analog has been described by Sonnenfeld (1996) and related to the Elk Basin Field. Exposure time is estimated at 10–20 million years, typical of a second order sequence boundary. Small-scale karst features are found capping the upper two high-frequency (third order) sequences. Anhydrite beds are common in much of the area but are absent in the area of the Elk Hills Field. This is attributed to efficient dissolution of anhydrite by undersaturated meteoric water in the Elk Hills area. The massive dissolution produced a strataform cave system which later collapsed, forming extensive strataform collapse breccia composed of a suite of breccia fabrics ranging from chaotic, to mosaic,

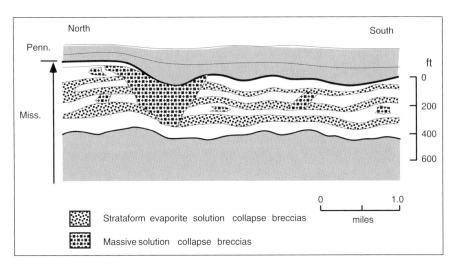

Fig. 10. Cross section of the Elk Basin (Mississippian) field, Wyoming, showing massive breccias that cross cut stratigraphy and strataform collapse breccias formed by dissolution of evaporite beds (After McCaleb and Waynam 1969)

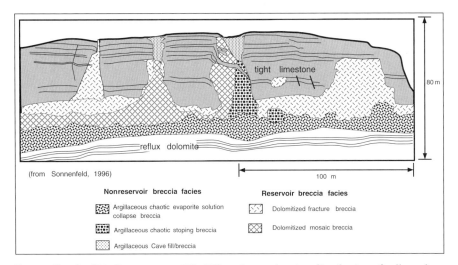

Fig. 11. Sketch of Madison outcrop, Wind River Gorge, showing distribution of collapse breccia. (Sonnenfeld 1996)

to crackle (Fig. 11). Cave collapse triggered a stoping process in overlying strata which propagated upward to the Mississippian unconformity. The vertical breccias are possibly guided by preexisting fractures spaced about 200 ft apart (Fig. 11). A late dolomitization event overprints the vertical breccias and mosaic and fracture breccia of the strataform roof facies.

Reservoir quality is concentrated in dolostone and collapse fractures of the roof facies. Limestone rarely has more than 2% porosity, and evaporite solution collapse breccias are flow barriers because they are argillaceous. Beds of peritidal sediments and associated reflux dolomite are typically porous and permeable, forming laterally continuous flow units. Late dolomitization controlled by collapse brecciation patterns also produces reservoir quality dolostones. However, their pattern will be nonstrataform and laterally discontinuous because of the irregular distribution of the collapse brecciation and fracturing.

7.3.3
Ellenburger Fields, West Texas

Ellenburger (lower Ordovician) reservoirs in the Permian Basin of West Texas and New Mexico are classified as fractured reservoirs, and the fractures are a product of massive dissolution and collapse. An outcrop analog to these reservoirs is found in the Franklin Mountains, El Paso, Texas. Major collapse breccias are associated with the second order sequence boundary between the Lower Ordovician El Paso Group and the upper Ordovician Montoya Group which has an estimated time gap of 30 million years. Extensive field mapping of the El Paso Group has demonstrated that a large cavern system was formed in the upper 300 m (1000 ft) of the El Paso Group (Fig. 12). In the upper 75 m (250 ft), the El Paso caverns were tabular and horizontal and were formed near the phreatic-vadose interface. These caverns were discontinuous but extended for thousands of feet laterally. In the lower 225 m (750 ft), the caverns were linear and vertical and were formed in the deep phreatic zone along vertical fractures spaced 900 m (3000 ft) apart. These caverns were also discontinuous, extending laterally along the precursor fractures for several kilometers and perpendicular to the fracture trend for hundreds of meters (Fig. 13). Collapse of the El Paso Caverns formed a cave roof fracture system in the overlying Montoya Group and megacollapse breccias in the El Paso Group. Collapse brecciation formed fractures in the overlying Ordovician and Silurian strata which were sites of later dissolution and collapse brecciation. A late dolomitization event converted much of the fractured and brecciated strata to dolomite, and dolomite cement filled much of the fracture and breccia pore space.

In the subsurface Ellenburger reservoirs, collapse breccias can be found as far as 1000 ft below the exposure surface at the top of the Ellenburger. The upper 300 ft, however, is of most interest because that is where most of the reservoirs are found for trap considerations. These reservoirs have been described by Kerans (1988, 1989) as having a fractured roof facies, a cave fill facies, and a cave floor breccia facies. Laterally persistent caverns have floor breccias formed during cave development that are overlain by clastic sediments of the Simpson formation that were washed into the resulting caves during transgression (figs. 14, 15, and 16). Later collapse produced extensive fracturing and brecciation, producing a vertical sequence of intact cave floor, cave floor breccia, cave fill breccia, and cave roof breccia.

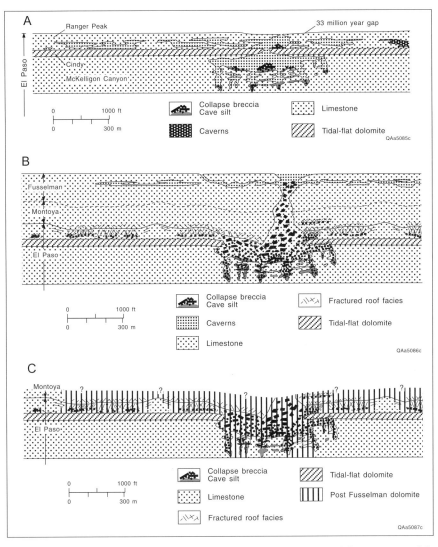

Fig. 12. Reconstruction of El Paso Caverns. **A** penecontemporaneous dolomitization of the Cindy Formation and development of tabular, laterally continuous caverns in the Ranger Peak Formation and vertical, laterally discontinuous caverns in the McKelligon Canyon Formation. **B** Collapse of the El Paso Caverns showing collapse of the Montoya, development of breccia pipes up into the Fusselman Formation, and development of caverns in the Fusselman Formation. **C** Late-stage dolomitization of the El Paso and Montoya Groups controlled by fluid flow through collapse breccia, fractures, and into adjacent carbonates. (Lucia 1995)

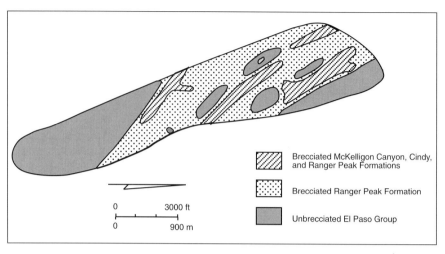

Fig. 13. Map view showing a reconstruction of the collapse breccia in the southern Franklin Mountains, El Paso, Texas. (Lucia 1995)

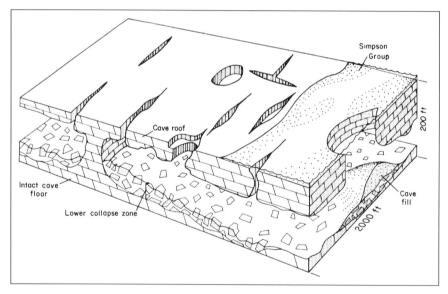

Fig. 14. Schematic block diagram showing a laterally extensive cave system with collapse breccias lining the cave floor. (Kerans 1989)

The cave fill breccia forms a reservoir barrier separating an upper and lower reservoir. SP and gamma ray logs are used to identify the cave filling clastic sediment and map the upper and lower reservoir. These fields have bottom water drives and initial development wells were completed in the upper reservoir only

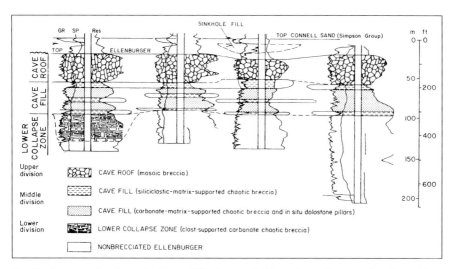

Fig. 15. Cross section from the Emma Ellenburger field, Andrews County, West Texas, illustrating the fractured cave-roof facies, laterally persistent cave fill facies, and laterally discontinuous collapse-breccia facies. (Kerans 1989)

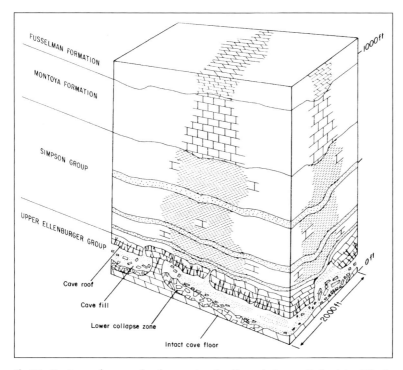

Fig. 16. Cartoon of cavern development and collapse in Lower Ordovician Ellenburger fields of West Texas showing cave facies (Kerans, 1989)

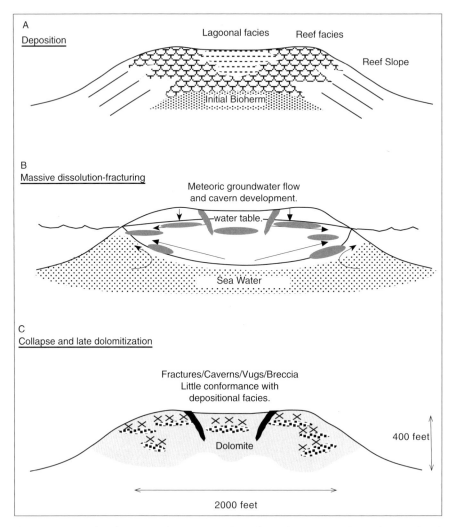

Fig. 17. Massive dissolution in Silurian pinnacle reefs, Michigan. **A** Reef growth started with a marine cemented bioherm on which corals and stromatoporoids grew and developed a reef margin, slope, and interior lagoon. **B** A freshwater lens developed during the subaerial exposure phase resulting in massive dissolution and the formation of caverns and associated enlarged fractures. **C** Collapse of the cave system with burial produced collapse breccias and fractures. (Modeled after Sears and Lucia 1980)

to control water coning. However, the recognition of a cave fill reservoir barrier and that water inflow was edge water, not bottom water (Ader 1980), initiated a redevelopment program to deepen wells to recover hydrocarbons in the lower reservoir.

7.3.4
Northern Michigan Silurian Reef Fields, Michigan

The Northern Michigan Reef belt produces from Silurian pinnacle reefs. The reefs are located parallel to and basinward of a massive shelf-edge reef complex. Reefs are found about every 1.5 square miles within the reef belt and average about 80 acres in size and 400 ft in height. The reefs were initiated in deep water with deposition of a mud mound facies and grew up into shallower water where corals and stromatoporoids formed the main constructional facies. A flanking crinoidal facies is present seaward and an algal restricted marine facies is found behind the coral-stromatoporoid facies. After it formed, the reef was exposed subaerially and massive dissolution occurred (Fig. 17). After deposition of thick salt beds adjacent to the reefs, the basin returned to normal marine conditions and a carbonate tidal flat sequence was deposited over and around the reefs. Dolomitization occurred at this time. With burial, salt-saturated waters were compacted out of the salt basin and into the reefs, occluding much of the porosity in the more basinward reefs with halite (Sears and Lucia 1980; Caughlin et al. 1976).

The massive dissolution is typical of small buildups and large sea level changes. The sea level changes were probably related to evaporative drawdown of the restricted Michigan Basin rather than to eustacy. A local rather than a regional groundwater system was developed within each buildup, and the associate massive dissolution formed caverns and enlarged fractures which were later dolomitized. Small bit drops that occurred while developing these reservoirs attest to the fact that some small caverns are still present.

References

Ader JC (1980) Stratification testing results in revised concept of reservoir drive mechanism, University Block 13 Ellenburger Field. J Pet Technol 32, 8: 1452–1458
Back W (1963) Preliminary results of a study of calcium carbonate saturation of a ground water in Central Florida: Int Assoc Sci Hydrol VIIIe, Annee 3, pp 43–51
Budd DA, Saller AH, Harris PA (eds) (1995) Unconformities and porosity in carbonate strata. AAPG Mem 63, 313 pp
Caughlin WG, Lucia FJ, McIver NL (1976) The detection and development of Silurian reefs in Northern Michigan. Geophysics 41, 4: 646–658
Choquette PW, James NP (1984) Introduction. In: James NP, Choquette PW (eds) Paleokarst. Springer, Berlin Heidelberg New York, pp 1–24
Dravis J J, Muir ID (1993) Deep-burial brecciation in the Devonian Upper Elk Point Group, Rainbow Basin, Alberta, Western Canada. In: Fritz RD, Wilson JL, Yurewicz DA (eds) Paleokarst related hydrocarbon reservoirs. SEPM Core Workshop 18, pp 119–167
Heydari E, Moore CH (1989) Burial diagenesis and thermochemical sulfate reduction, Smackover Formation, southeastern Mississippi salt basin. Geology 17, 12: 1080–1084
Kaufman J, Meyers WJ, Hanson GN (1990) Burial cementation in the Swan Hills Formation (Devonian), Rosevear Field, Alberta, Canada. J Sediment Petrol 60, 6: 918–939
Kerans C, Lucia FJ, Senger RK (1994) Integrated characterization of carbonate ramp reservoirs using outcrop analogs. AAPG Bull 78, 2: 181–216
Kerans C (1988) Karst controlled reservoir heterogeneity in Ellenburger Group carbonates of west Texas. AAPG Bull 72, 10: 1160–1183

Kerans C (1989) Karst-controlled reservoir heterogeneity and an example from the Ellenburger Group (Lower Ordovician) of West Texas. The University of Texas at Austin, Bureau of Economic Geology, Report of Investigations No 186, 40 pp

Loucks RG, Handford CR (1992) Origin and recognition of fractures, breccias, and sediment fills in paleocave-reservoir networks. In: Candelaria MP, Reed CL (eds) Paleokarst, karst-related diagenesis and reservoir development: examples from Ordovician-Devonian age strata of West Texas and the Mid-Continent. Permian Basin Section-SEPM Publ 92-33, Midland, Texas, pp 31-44

Lucia FJ (1970) Lower Paleozoic history of the western Diablo Platform of West Texas and south central New Mexico. Geologic Framework of the Chihuahua Tectonic Belt. West Texas Geol Soc Publ, Midland, Texas, pp 39-55

Lucia FJ, Kerans C, Vander Stoep GW (1992) Characterization of a karsted, high-energy, ramp-margin carbonate reservoir: Taylor-link West San Andres Unit, Pecos County, Texas. The University of Texas at Austin, Bureau of Economic Geology, Report of Investigation 208, 46 pp

Lucia FJ (1995) Lower Paleozoic development, collapse, and dolomitization, Franklin Mountains, El Paso, Texas. In: Budd DA, Saller AH, Harris PA (eds) Unconformities and porosity in carbonate strata. AAPG Mem 63: 279-300

Maiklem WR (1971) Evaporative drawdown - a mechanism for water-level lowering and diagenesis in the Elk Point Basin. Bull Can Petrol Geol 19, 2: 487-503

McCaleb JA, Wayhan DA (1969) Geologic reservoir analysis, Mississippian Madison Formation, Elk Basin Field, Wyoming, Montana. AAPG Bull 51: 2122-2132

Plummer FN (1975) Mixing of sea water and calcium carbonate ground water: In: Whitten EHT (ed) Quantitative Studies in Geological Sciences. GSA Mem 112: 219-236

Sears SO, Lucia FJ (1980) Dolomitization of Northern Michigan Niagaran reefs by brine refluxion and freshwater/seawater mixing. In: Zenger DH, Dunham, JB, Ethington RL (eds) Concepts and models of dolomitization. SEPM Spec Publ 28: 215-236

Sonnenfeld MD (1996) An integrated sequence stratigraphic approach to reservoir characterization of the Lower Mississippian Madison Limestone, emphasizing Elk Basin Field, Bighorn Basin, Montana. Unpubl PhD Dissertation, Colorado School of Mines, Golden, Colorado, 438 pp

Stoessell RK, Ward WC, Ford BH, Schuffert JD (1989) Water chemistry and CaCO3 dissolution in the saline part of an open-flow mixing zone, coastal Yucatan Peninsula, Mexico. GSA Bull 101, 2: 159-169

Tinker SW, and Mruk DH (1995) Reservoir characterization of a Permian giant: Yates field, West Texas. In: Stoudt EL, Harris PM (eds) Hydrocarbon reservoir characterization: Geologic framework and flow unit modeling. SEPM Short Course 35: 51-128

CHAPTER 8

Reservoir Models for Input into Flow Simulators

8.1
Introduction

Reservoir characterization is defined as the construction of realistic three-dimensional images of petrophysical properties to be used to predict reservoir performance. A key element in constructing reservoir models is modeling the high and low permeabilities. In the past, these images have been prepared by several different methods including (1) the layered reservoir method, (2) the continuous pay method, and (3) the facies method (Fig. 1). In the layered reservoir method, the reservoir is divided into pay zones using correlations based on gamma ray logs, and the net feet of porosity (net pay maps) or the porosity-times-oil-saturation (SoPhiH maps) isopached for each layer. The layers commonly lump several petrophysical rock types, and the average petrophysical values do not characterize the flow properties of the reservoir.

In the continuous pay method, the layers are defined as porous intervals and are assumed to correlate laterally parallel to structure. Porous intervals are extended horizontally between wells and the percentage of porous intervals that are continuous between any two wells calculated. The expected incremental recovery of an infill well is inversely related to the percent continuity of the individual porous intervals (George and Stiles, 1978; Barber et al. 1983). This method is best applied in reservoirs with discrete reservoir and nonreservoir intervals.

In the facies method, the reservoir is layered using correlatable gamma-ray markers, depositional facies are described from core material and mapped within the layers, and average petrophysical properties are assigned to each facies. Depositional facies are commonly defined by their fossil content and other depositional features for the purpose of predicting and mapping facies distribution. This is a useful method for delineating reservoir patterns and is most effective in exploration and initial field development. It is not effective in reservoir characterization because 1) the layers are not chronostratigraphic and 2) geological facies commonly lump several rock fabrics so that the average petrophysical properties do not characterize fluid flow in the reservoir. To be useful, the facies must be defined in terms of petrophysical rock fabrics, as described in Chapter 2.

In previous chapters we have discussed petrophysical properties and emphasized that permeability and saturation are a function of pore-size distribution

Fig. 1. Three methods of reservoir characterization. **A**
the layered reservoir method: **B** the continuous pay
method: **C** the facies method

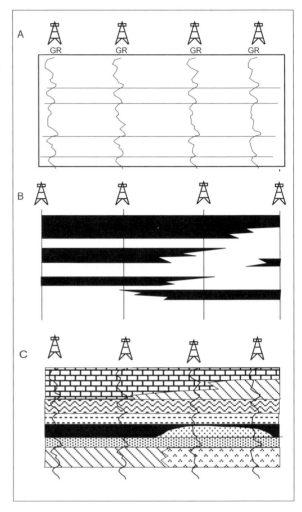

Fig. 1. Three methods of reservoir characterization. **A** the layered reservoir method: **B** the continuous pay method: **C** the facies method

and that pore-size distribution is a function of particle size, sorting, interparticle porosity, separate-vug porosity, and touching-vug pore systems. We have discussed the use of core description and wireline logs for characterizing the one-
dimensional distribution of the rock fabrics and associated petrophysical values.
The distribution of rock-fabric facies within a chronostratigraphic framework
has been emphasized as the proper method of constructing a geological model
suitable for petrophysical quantification. The effect of diagenesis on the distribution of petrophysical properties has been discussed in terms of how diagenesis affects our ability to image the distribution of petrophysical properties.

   We now have the tools to construct a geological model and are ready to construct a reservoir model suitable for input into fluid flow simulators. The accuracy of the reservoir model will depend upon the accuracy of the geological
model, which in turn depends in part on the diagenetic environment. There is a

wide range of predictability, and we will discuss those reservoirs where the diagenetic products can be closely tied to depositional fabrics.

Three basic steps in the construction of a reservoir model are (1) constructing the 3-D geological framework and mapping the rock-fabric facies within each layer, (2) estimating the one-dimensional distribution of porosity, permeability, and fluid saturation at each well, and (3) filling the framework with petrophysical properties. The first two steps have been discussed in previous chapters. In this chapter we will discuss the difficult task of filling the geologic framework with petrophysical properties using one-dimensional well data.

The basic methods used to distribute petrophysical data within the stratigraphic framework include mapping rock-fabric facies, linear interpolation between wells, geostatistical variography, and stochastic distribution of petrophysical objects. In this chapter we will focus on the rock-fabric method of constructing a reservoir model. We will refer to geostatistical data, and a brief overview of variography is presented before discussing the rock-fabric method. No discussion of stochastic methods to distribute facies is included because of the lack of data on the dimensions of carbonate facies.

8.2
Geostatistical Methods

8.2.1
Introduction

Geostatistical methods focus on describing heterogeneities that are on a scale smaller than well spacing. The approach is to estimate interwell permeability distribution on the basis of detailed vertical permeability profiles calculated from wireline logs and a stochastic geostatistical method to model the interwell heterogeneity that honors the vertical profiles.

8.2.2
Variograms

The following is taken from Fogg and Lucia (1990). Modern geostatistics differs from classical population statistics in that geostatistics is designed to handle data that exhibit spatial correlation; that is, the data are not totally random, and neighboring values bear some relation to each other. Nearly all geology-related data exhibit spatial correlation. The likelihood that measurements of a spatially correlated variable are close or equal in value increases as distance between the points of measurement decreases. Another distinction between the two methods is that in classical statistics, the mean and variance are generally the fundamental measures of a distribution, whereas in geostatistics, the mean and covariance or variogram are the fundamental measures. The covariance and variogram are statistical functions used to describe spatial correlation.

In reservoir characterization, geostatistical methods are used to generate realistically complex interwell patterns of heterogeneity, which are critical for recovery efficiency. These methods are also used to estimate probabilities for various interwell permeability patterns and, in turn, to estimate error bars for simulated production and recovery efficiency. The geostatistical approach is essentially a quantitative geological mapping technique in which geological interpretation is included through the variogram. Geostatistics in no way lessens the need for valuable, subjective mapping by the geologist. On the contrary, properly applied geostatistics increases the need for geological input and helps define which geological data are most applicable to a particular problem.

8.2.2.1
Variography

Variography uses variograms to statistically characterize spatial variability of a property. In essence, the variogram is a tool that under certain circumstances can be used to quantify the spatial continuity that is either explicitly exhibited in the data or inferred from the data through geological interpretation.

To visualize how the variogram can be estimated from data, consider the data points along the line in Fig. 2. At each point, the variable of interest, Y (Y could equal permeability), is known. Each pair of numbers can be grouped into a distance class, depending on the distance separating the pair. An experimental variogram value (γ) can then be calculated for each distance class from the mean of the squared differences between pairs of values (see equation in Fig. 2).

When (γ) is calculated as a function of distance for data that are spatially correlated, the values commonly form a curve like the one in Fig. 2. The value of (γ) rises from near zero and then levels off into a segment known as the sill. When

Fig. 2. Schematic example showing how the experimental variogram can be calculated in one dimension (Fogg and Lucia 1990)

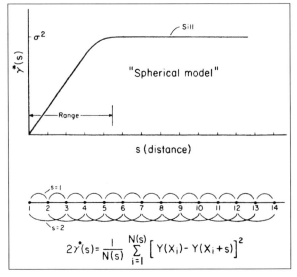

$$2\overset{*}{\gamma}(s) = \frac{1}{N(s)} \sum_{i=1}^{N(s)} \left[Y(X_i) - Y(X_i + s) \right]^2$$

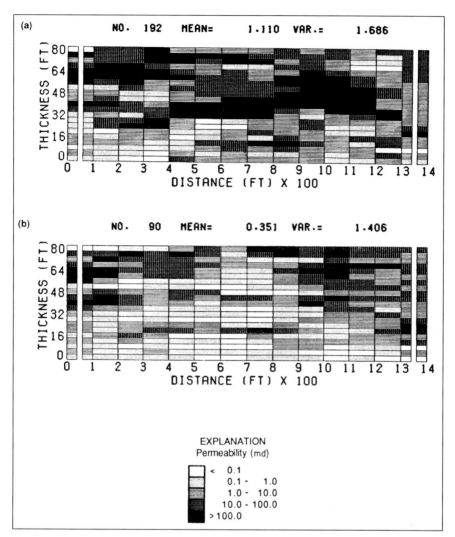

Fig. 3. Spatial distribution of permeability in two stochastic realizations. **a** Realization representing relatively high permeability and continuity. **b** Realization representing relatively low permeability and continuity (Fogg and Lucia 1990)

a horizontal sill exists, the value of (γ) at the sill is equal to the variance of Y. The curve in Fig. 2 indicates that the variability, or variance, of the data pairs increases with distance until the separation distance is too large for values to be correlated. The distance at which (γ) stops increasing significantly is known as the range. The initial value of (γ) may not be zero, but may be any value up to the value of (γ). This is known as the nugget effect, and the higher the initial value is, the poorer is the spatial correlation at the small scale.

8.2.2.2
Conditional Simulation

Conditional simulation reproduces or simulates the true variability of a field such that the simulated values (1) vary stochastically between data points as a function of the variogram and data distribution and (2) honor the data points. The simulation produces the same degree of variability and spatial correlation as the variogram and honors all data points. Out of hundreds of realizations of a conditional simulation, one or more realizations would probably approximate the true distribution. Conditional simulation, however, is not used to estimate reality but to produce realizations that have the same degree of spatial variability and complexity as reality. Spatial distributions of permeability in two stochastic realizations illustrated in Fogg and Lucia (1990) are shown in Fig. 3. Realization (a) represents relatively high permeability and continuity, and (b) represents relatively low permeability and continuity.

8.3
Scale of Variability and Average Properties

A characteristic of carbonate reservoirs is extreme scatter when porosity and permeability are graphed (Fig. 4). We have identified some of the reasons for this scatter through our rock fabric classification. In nonvuggy carbonates the scatter is reduced by grouping data by particle size and sorting. In carbonates with separate vugs the scatter within a rock-fabric class can be reduced by grouping the data into bins with equal separate-vug porosity. However, the porosity and permeability cross-plots do not address the problem of spatial distribution, and we need to address this question next.

In previous chapters we have distributed petrophysical properties by assigning values to rock-fabric facies. However, there is considerable variability in petrophysical properties within a rock-fabric facies. This is illustrated by permeability profiles from vertical traverses taken from an outcrop of dolograinstone (Fig. 5A). The vertical variability in each profile is commonly assumed to be correlatable between the profiles, and correlation is accomplished by using pattern recognition and the "law of horizontality". The result is a layered permeability model (Fig. 5B). Because this is an outcrop, however, we can collect detailed permeability data from between the vertical profiles and find that the permeability is not layered but is characterized by "bullseyes" or a patchy permeability pattern (Fig. 6).

The question of the spatial distribution of permeability and porosity within a rock-fabric facies has been examined by making extensive permeability measurements on the San Andres outcrop found in Lawyer Canyon, Algerita Canyon, Guadalupe Mountains, New Mexico (Senger et al. 1993; Grant et al. 1994). Permeability measurements were made on the scale of inches to feet in vertical transects, horizontal transects, and grids. One set of data is concentrated in the dolograinstone facies of cycle 1. Analysis of the data shows that the scale of var-

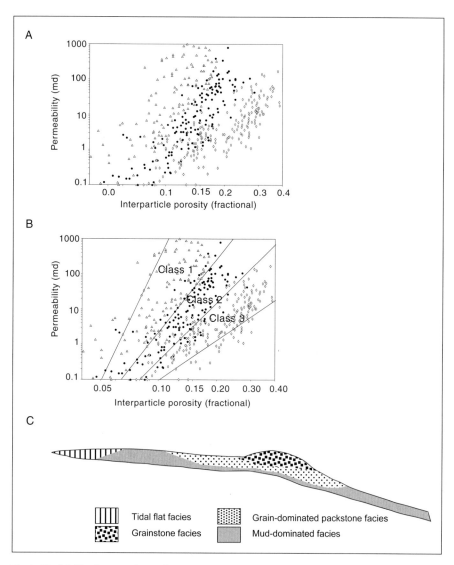

Fig. 4. Variability in porosity and permeability in carbonate rocks. A Porosity-permeability cross plot showing extreme scatter typical of carbonate reservoirs. B Systematic organization of variability into petrophysical classes based on rock fabrics and interparticle porosity. C Systematic spatial distribution of rock-fabric facies within a depositional cycle

iability is a function of the sample spacing. Permeability was sampled every foot in vertical traverses spaced 50 to 150 ft along a 2600-ft lateral distance (Fig. 7a), every foot in vertical traverses spaced every 5 ft over a lateral distance of 40 ft (Fig. 7b), and every foot in a 20 x 15 ft grid pattern (Fig. 7c). Permeability iso-

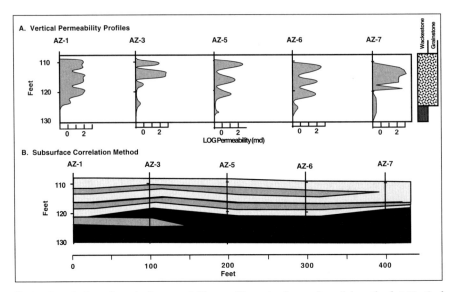

Fig. 5. Correlation of vertical permeability profiles using layered model method. A Vertical permeability data from outcrop measured sections. B Correlation of permeability based on layered model

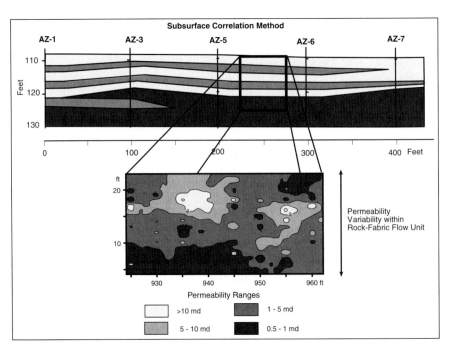

Fig. 6. Small-scale variability within a rock fabric strata compared with layered permeability model. Permeability structure is not layered but consists of a series of uncorrelated closed isopachs

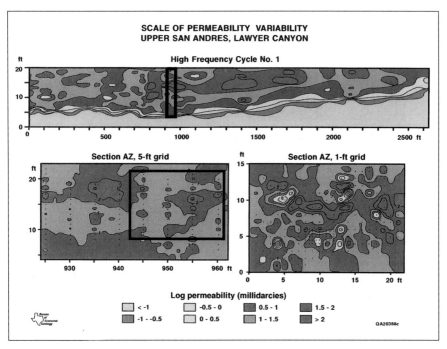

Fig. 7. Permeability distribution in cycle 1 grainstone rock-fabric facies at three different scales. A 20 by 2500 ft scale with 1ft data from vertical transects spaced 50 to 150 ft apart. B 17 by 35 ft scale with 1ft from vertical transects spaced 5 ft apart. C 15 by 20 ft scale with 1ft data from vertical transects spaced 1ft apart.

pach maps were made for each case, and a comparison of the three maps shows variability on the scale of several orders of magnitude with the scale of variability related to the sampling density. Extensive geostatistical analyses have been made on this data (Senger et al. 1993; Grant et al. 1994) and the resulting variograms typically show a high nugget, indicating high variability at the small scale and a near-random spatial distribution (Fig. 8). *Therefore, within a rock-fabric facies, permeability varies several orders of magnitude on the scale of feet to inches and the poor spatial correlation suggests that the data can be legitimately averaged for modeling purposes.*

Simulation experiments were conducted on the grainstone facies to examine the effect of using various permeability distributions on simulation results (Senger et al. 1993). Comparisons between recovery efficiency using a spatially distributed permeability structure and a single average value show little difference, suggesting that permeability values can be legitimately averaged on the scale of rock-fabric units (Fig. 9). The production rate, however, was higher for the statistical distribution than for the average value. This effect was also found by Grant et al. (1994), and they also show that injection rates are higher when a variogram is used to stochastically distribute permeability data.

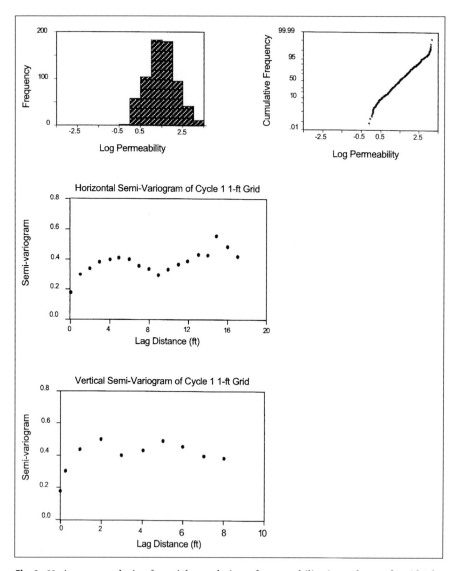

Fig. 8. Variogram analysis of spatial correlation of permeability in cycle 1, 1-ft grid. The high nugget indicates poor spatial correlation of permeability data at the small scale

Therefore, *a flow unit can be defined as a geologic unit that has a characteristic rock fabric, and within which the petrophysical properties are near randomly distributed* (Lucia et al. 1992). This definition amplifies the flow unit definition of Hearn et al. (1984). They define flow units as representing fairly distinctive rock types and having petrophysical characteristics significantly different from those of adjoining units.

Fig. 9. Performance predictions from flow simulations of cycle 1 grainstone. A Comparison of water flood recovery using various methods for representing permeability distribution in cycle 1 grainstone showing little difference between stochastic and facies-averaging methods. **B** Comparison of oil production rate using various methods for representing permeability distribution in cycle 1 grainstone showing higher initial production rates for stochastic simulations than facies-averaged simulations (Senger et al. 1993)

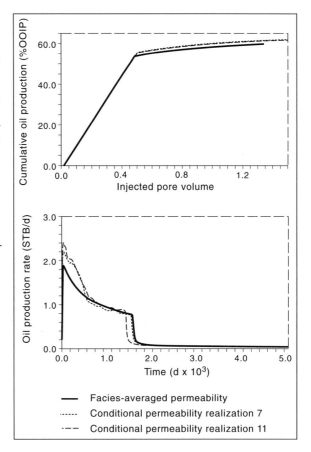

Subsurface studies show high variability of porosity and permeability at the scale of inches and feet similar to that observed in the outcrop. A core from well 2505, Seminole San Andres Unit, Gaines County, Texas, was used to study permeability variability. Twelve even-textured samples for which whole-core permeability values were available were selected. The samples include dolograin-stone, mud-dominated dolopackstone, and dolowackestone. A 1-in. permeability grid of the 7- by 3.5-in. slab surface of the 12 samples was measured using a mechanical field permeameter (MFP; Fig. 10). Three 1-in. plugs were then drilled from the whole-core samples, analyzed for porosity and permeability, and the permeability of each end of the plug measured with a MFP.

The permeability values were compared with the whole core permeability, and examples of the results are presented in Figs. 11, 12, and 13. Permeability varies between 0.5 and 2 orders of magnitude on the scale of inches, that is, within each core sample (Fig. 14). Neither the geometric nor the arithmetic mean values match the whole-core permeability. Whole-core permeability is best characterized by the geometric mean in 50% of the samples and the arithmetic

Fig. 10. Sampling method for detailed permeability study, Seminole San Andres Unit well 2505

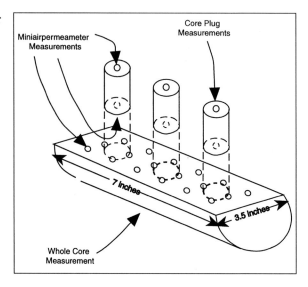

Fig. 11. Detailed permeability results from sample 5106 ft, Seminole San Andres Unit well 2505 showing variability factor of 10

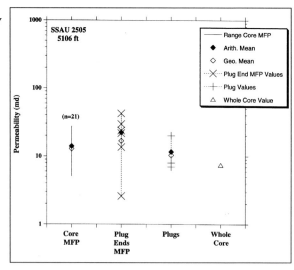

mean in 50% of the samples, suggesting that, at this scale, any average value between the geometric and arithmetic mean can be considered appropriate.

8.4
Rock-Fabric Reservoir Models

Reservoir models should be realistic images of the three-dimensional distribution of petrophysical properties, and it is generally agreed that the most important values to image properly are high and low permeability values. High perme-

Fig. 12. Detailed permeability results from sample 5143 ft, Seminole San Andres Unit well 2505 showing variability factor of 100

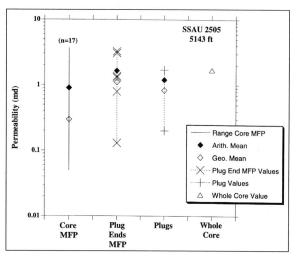

Fig. 13. Detailed permeability results from sample 5167 ft, Seminole San Andres Unit well 2505 showing variability factor of 50

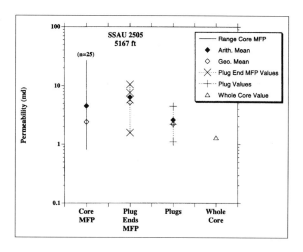

ability layers control water breakthrough and patterns of bypassed oil whereas low permeability values tend to be baffles and inhibit cross flow. More realistic reservoir models can be constructed from outcrop data because detailed two- and three-dimensional geological and petrophysical data can be obtained as opposed to one-dimensional data from subsurface reservoirs. The outcrop study at Lawyer Canyon on the Algerita Escarpment, Guadalupe Mountains, New Mexico provides an opportunity to construct a reservoir model of a carbonate ramp reservoir and investigate the impact of various geological elements on performance prediction.

Fig. 14. Summary of results of detailed permeability study from 12 core samples, Seminole San Andres Unit well 2505. Variability within a core sample is of the order of 1 or 2 orders of magnitude (variability factor of 10 to 100). Assuming the whole-core value to be ground truth permeability, neither the geometric or arithmetic mean values consistently match the true average permeability

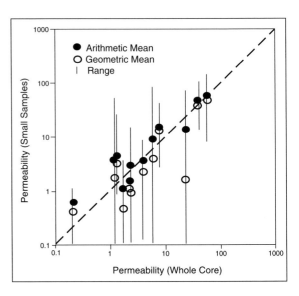

8.4.1
Lawyer Canyon Reservoir Analog Study

8.4.1.1
Model Construction

The Lawyer Canyon reservoir window is a two-dimensional outcrop composed of nine upward-shallowing depositional cycles (Fig. 15). The cycles are typically upward-shallowing subtidal cycles composed of a vertical sequence of dolomudstone at the base, grading upward through dolowackestone and dolopackstone to cross-bedded dolograinstone at the top. Locally, the cycles are capped by a fenestral fabric interpreted to represent shoaling into the intertidal environment. The sequence is not always complete, but a vertical increase in grain size and sorting can typically be observed within any one cycle.

Cycles 1 and 2 are dolograinstone-capped cycles. Cycle 1 has a continuous basal mudstone whereas cycle 2 has discontinuous thin dolomudstone beds at the base. Cycles 3-6 are primarily composed of grain-dominated dolopackstone in the southern area and dolowackestone in the northern area of the study window. Cycle 3 contains a small ooid dolograinstone body in the southern area, but cycles 4-6 are devoid of dolograinstone. Discontinuous fenestral beds are found capping some cycles, and discontinuous tight and dense dolomudstones are found at the base of the cycles.

Cycle 7 is a dolograinstone-capped cycle and is unique in that it contains large amounts of moldic pore space (separate vugs). Cycle 8 is a thin moldic grain-dominated dolopackstone or grainstone. Cycle 9 contains an upper dolograinstone unit that varies in thickness from 0 to 10 ft in the north, to 40 ft in the south and terminates in the southern portion of the study area.

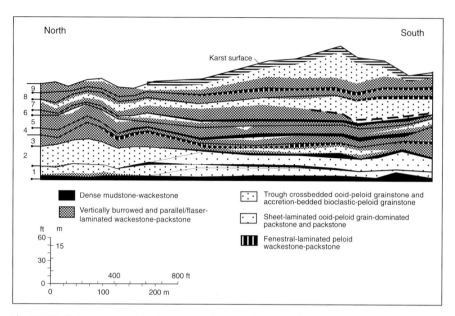

Fig. 15. High frequency cycle framework, Lawyer Canyon, Algerita Escarpment (Kerans et al. 1994)

The interval has been completely dolomitized, probably from refluxing hypersaline marine water from a source in the overlying tidal flat and evaporite beds. Dolomitization occurred soon after deposition and shallow compaction/cementation and resulted in a high conformance between depositional texture and rock fabric. The moldic porosity in cycle 7 is an example of selective dissolution associated with a local freshwater lens (Hovorka et al. 1993)

The high-frequency cycles are composed of five fundamental rock fabrics: (1) dolograinstone, (2) moldic dolograinstone, (3) grain-dominated dolopackstone, (4) fine crystalline mud-dominated dolopackstone/wackestones, and (5) dense, fine crystalline dolowackestone/mudstone. Core plugs taken from each of these rock fabrics were analyzed for porosity and permeability. The cross plot of interparticle-porosity and permeability exclusive of the moldic dolograinstone shows that the rock fabrics fall into the generic rock fabric/petrophysical fields (Fig. 16A). A cross plot of total-porosity and permeability for samples from the moldic grainstone facies shows a general grouping by moldic porosity (Fig. 16B).

A flow model was constructed of the Lawyer Canyon window using the rock-fabric method. A chronostratigraphic framework composed of nine high-frequency cycles was established by defining high-frequency cycles from upward-shallowing fabric successions. The rock-fabric facies were mapped within the cycle framework and porosity and permeability values from each rock-fabric facies were averaged using an arithmetic average for porosity and geometric average for permeability. Water saturation values were obtained from generic

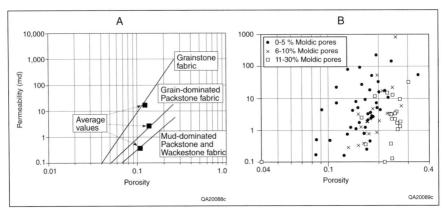

Fig. 16. Porosity/permeability relationships for **A** nonvuggy rock-fabric types and **B** separate-vug (moldic) pore types. The nonvuggy pore types have rock-fabric specific transforms. The moldic grainstones do not have a transform but can be related to the grainstone transform by subtracting moldic porosity from total porosity (Kerans et al. 1994)

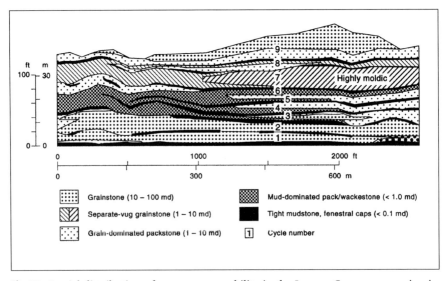

Fig. 17. Spatial distribution of average permeability in the Lawyer Canyon reservoir window. Permeability is averaged within rock-fabric units (Lucia et al. 1992)

rock-fabric-specific Sw, porosity, and reservoir-height relationships discussed in Chapter 2.

Fig. 17 is a generalized illustration of the Lawyer Canyon rock-fabric reservoir model. Key elements are the high-permeability grainstones of cycles 1, 2, and 9, the low-permeability discontinuous mudstones at the base of most cycles, and the high-porosity low-permeability moldic grainstone of cycle 7. The cycle boundaries and distribution of rock fabrics are known to a high degree of certainty because they have been walked out and mapped in the field. The average petrophysical data of some rock-fabric facies are less certain because of the small amount of data.

8.4.1.2
Fluid Flow Experiments

Flow simulation experiments using the Lawyer Canyon reservoir model were conducted to test the sensitivity of performance prediction to certain geological elements. The results of these flow experiments illustrate that (1) the inclusion of realistic geological and petrophysical heterogeneity is critical for obtaining realistic performance predictions from simulators and (2) accurate perform-ance prediction requires careful integration of reservoir heterogeneity (static) and operational procedures (dynamic). Simulated oil recovery from waterflood-ing using (1) the outcrop reservoir model and (2) a simple layer model con-structed by interpolation of geologic and petrophysical data between the two ends of the Lawyer Canyon window show a recovery of almost 50% of the oil in place using the layered model compared with about 35% recovery using the ground-truth outcrop model (Figs.18 and 19).

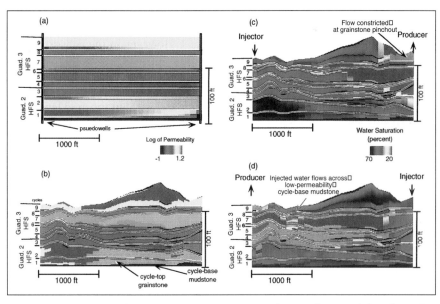

Fig. 18. Lawyer Canyon flow models. **a** A linear interpolation of permeability data taken from two pseudowells on each end of the Lawyer Canyon window. **b** The rock-fabric perme-ability model based on facies averaging of outcrop data. **c** Left-to-right injection experiment showing water saturation after 40 years of injection and crossflow point at downflow termi-nation of high permeability in cycle 9. **d** Right-to-left injection experiment showing water saturation after 40 years of injection and crossflow point at downflow termination of high permeability in cycle 9 (Lucia et al. 1995)

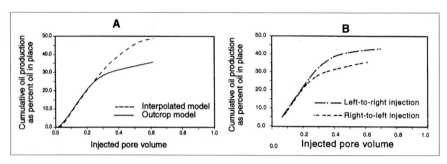

Fig. 19. Sweep efficiency plots comparing **A** linear interpolation model with outcrop model and **B** left-to-right with right-to-left injection experiment (Lucia et al. 1995)

Waterflood simulation experiments using the Lawyer Canyon model show a large difference in oil recovery between flooding in different directions. Oil recovery by flooding left to right is 44%, whereas flooding right to left recovers only 35%. This difference can be related to the distance between the producer and the termination of the high-permeability grainstone in cycle 9. In both simulations, cross-flow trapping of oil saturation in cycle 7 is produced by cross flow from cycle 9 into cycle 7 at the down-stream termination of the permeable grainstone in cycle 9, and oil recovery is inversely related to the distance between the producer and the grainstone termination (Figs. 18 and 19).

In all the flow models the kv/kh was set at 1. However, in the subsurface the kv/kv ratio often must be set at less than 1 (commonly 0.01 or less) for predicted and historical performance to match. The Lawyer Canyon model contains thin discontinuous dense mud layers that act as baffles to fluid flow, restricting gravity flow. Flow experiments were conducted with and without the mud layers and with various values of kv/kh (Fig. 20). The results showed that a kv/kh of 0.01 was necessary in the model without mud layers to match recovery from the model with mud layers and a kv/kh of 0.5. Therefore, the inclusion of thin dense layers in carbonate flow models is necessary to simulate realistic flow performance.

8.5
Reservoir Model Construction using the Rock-Fabric Facies Method

Outcrop studies suggest that the basic building block for constructing carbonate ramp reservoir models is the rock-fabric facies, and that these facies are systematically distributed within high-frequency cycles and sequences. Cycle and sequence boundaries are time surfaces and, therefore, continuous from well to well. They form a deterministic framework with the only uncertainty being the location of the correlative surface in any particular well. Rock-fabric facies commonly have abrupt vertical boundaries that can be defined in a well with varying degrees of accuracy and carried laterally, guided by the cycle boundary. Lateral dimensions, however, are of unknown length and must be estimated geostatisti-

Fig. 20. Waterflood recovery as a function of pore volume injected for Kv/Kh ratio of 0.001, 0.01, 0.1 and 1 using Lawyer Canyon outcrop model **(a)** with dense mudstone layers as baffles to gravity flow, and **(b)** without dense mudstone layers. **(c)** A Kv/Kh ratio of 0.3 is realistic based on core data and a ratio of 0.02 is required to compensate for the lack of baffles in the model without dense mudstone layers (Wang et al. 1998)

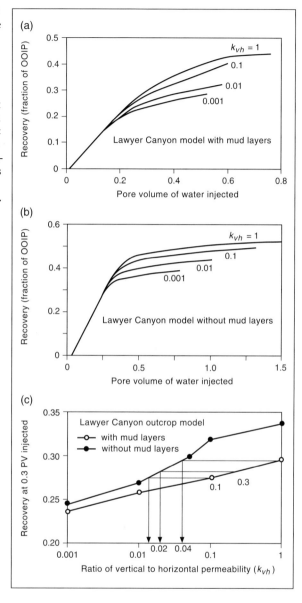

cally. Lateral dimensions are usually estimated based on depositional models because there is little formalized data on the dimensions of carbonate facies to use stochastic geostatistical methods.

Lateral boundaries of rock-fabric facies are not sharp; they are gradational from one facies to another over distances of hundreds of feet when observed in outcrops. Therefore, petrophysical properties are expected to change abruptly in the vertical dimension and gradually in the lateral dimension. This suggests that petrophysical properties can be linearly interpolated between wells in reservoirs with closely spaced well control.

8.5.1
Construction of Rock-Fabric Layers

A deterministic/statistical method of constructing a flow model suitable for simulation studies can be employed to distribute rock-fabric facies within the geological framework in reservoirs with closely spaced well control. The approach is to construct rock-fabric layers because most reservoir simulation programs are designed to handle layered data. Each layer may contain more than one rock-fabric facies, and the lateral boundary between these facies will be gradational. There are four steps in the construction of a rock-fabric layered model; (1) construction of the HFC framework, (2) identification of petrophysically significant rock-fabric facies within each well, (3) construction of rock-fabric layers, and (4) averaging petrophysical properties within rock fabric facies and linear interpolation of properties between wells.

The first step is to identify chronostratigraphic surfaces in each well using vertical facies succession, and to correlation these time surfaces from well to well (Fig. 21 A). The second step is to identify the petrophysical classes in each well and establish the proper permeability and saturation transforms to be used along with porosity and reservoir height to estimate permeability and original oil saturation (Fig. 21B). These first two steps should be done in parallel because they are the basic level at which geological interpretations and engineering data are integrated. The vertical facies succession, rock-fabric facies, and petrophysical classes must be linked before constructing rock-fabric layers.

Once the rock-facies have been described and the chronostratigraphic framework constructed, the next step is to construct rock-fabric layers (Fig. 21 C). Cycle boundaries will define a rock-fabric layer, and the vertical boundary of each rock-fabric facies in each well will also define a rock-fabric layer. The recommended method is to first pick the tops of rock-fabrics in each well and then correlate these laterally. Because rock-fabric layers must be continuous to be suitable for input into many reservoir simulators, there will be more rock-fabric layers than facies in some wells, and the layers must be carried through at some arbitrary point. This will result in some rock-fabric facies with more than one rock-fabric layer, but each rock-fabric layer will have the same average petrophysical properties. In the last step, the petrophysical properties at each well are averaged within rock-fabric intervals using the petrophysical classes to determine porosity/permeability transforms. The average values are interpolated between wells within the rock-fabric layers (Fig. 21D). Petrophysical properties can be distributed within the rock-fabric layers between wells by variography conditioned on well data and by interpolation of well data between wells. The advantage of variography over simple interpolation is that the continuity of permeability within each rock-fabric layer can be adjusted by using various variograms to match production and injection rates. However, whereas vertical vari-

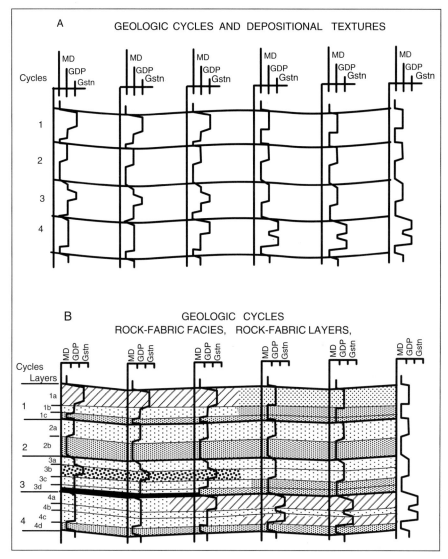

Fig. 21. Diagram illustrating steps for constructing a reservoir model using the rock-fabric method. **A** Construction of high-frequency sequence framework. **B** Identification of petrophysically significant rock-fabric facies and construction of rock-fabric layers.

ograms can be readily calculated from well data, the critical horizontal variograms needed to distribute permeability between wells must be guessed at or estimated from outcrop data.

An important test of the accuracy of any reservoir model is how closely the image reflects the geological model. Detailed outcrop studies provide most of

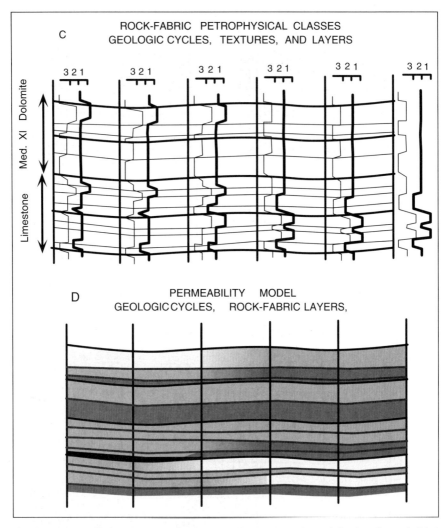

Fig. 21. Diagram illustrating steps for constructing a reservoir model using the rock fabric method (continued). **C** Identification of petrophysical classes for calculating saturations and permeability from wireline logs. **D** Interpolation of rock-fabric averaged petrophysical properties between wells

our concepts of how a reservoir should look. In the next section we will look at a field study of a carbonate ramp field that is very similar to the Lawyer Canyon outcrop study, and the reservoir model should reflect an architecture of high-frequency cycles similar to the Lawyer Canyon rock-fabric model.

Four types of models are illustrated in Fig. 22. The first illustration is the rock-fabric model constructed using the approach described above. The second

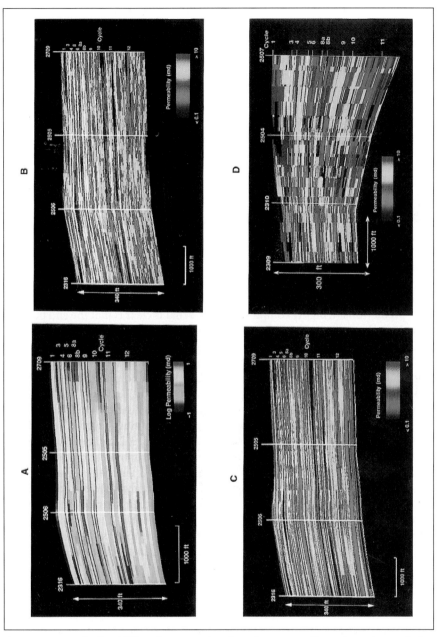

Fig. 22. Four methods of filling the reservoir volume between wells. A Linear interpolation of well data averaged within rock-fabric units and constrained by cycle boundaries. B Stochastic distribution of permeability based on variograms and constrained by rock-fabric units. C Linear interpolation of point well data constrained by cycle boundaries. D Stochastic distribution of permeability based on vertical and horizontal variograms but not constrained by cycle boundaries

is a geocelluar model with 1-ft permeability data linearly interpolated between wells parallel to the cycle boundaries. This approach is not realistic because outcrop data shows that permeability is near randomly distributed on the rock-fabric scale and not laterally correlatable on a 1-ft scale. The third model is the rock fabric model with permeability distributed stochastically within the rock-fabric flow units, using vertical variograms from well data and estimated horizontal variograms. This image is perhaps the most realistic and results in higher production and injection rates than the linearly interpolated rock-fabric model. The last model is a stochastic model where permeability is distributed without regard to the high-frequency cycles. The permeability patterns show no resemblance to the cycle architecture of the geological model.

8.6
Field Example: Seminole San Andres Unit, Gaines County, Texas

8.6.1
Introduction

Characterization of a carbonate reservoir for fluid flow simulation and performance prediction is a highly complicated task. What is clear is that the end product must be a three-dimensional numerical image of petrophysical properties: porosity, fluid saturation, permeability, and relative permeability. The principal problems are (1) obtaining the proper petrophysical values to be imaged and (2) distributing petrophysical values in 3-D space. We have discussed petrophysical properties, their dependence on pore-size distribution, and how pore-size distribution is linked to rock-fabrics. A one-dimensional distribution of rock fabric descriptions and petrophysical properties can be obtained from core material and wireline logs, providing data for distribution in 3-D space. We have discussed geological concepts used to distribute the petrophysical properties in 3-D space (chronostratigraphic surfaces, depositional facies progression, and diagenetic overprints), and we have suggested that rock fabric facies (depositional textures plus diagenetic overprint) can be systematically distributed within the chronostratigraphic framework if the diagenetic overprint conforms to depositional textures. In this chapter, we have shown that petrophysical properties are near randomly distributed within rock-fabric facies, and that a reservoir model constructed of rock-fabric layers can be populated with petrophysical data by averaging data within a rock-fabric and linear interpolation between wells.

The rock-fabric method of model construction consists of (1) measuring petrophysical properties on core material and relating the properties to rock fabrics, (2) describing the vertical succession of depositional textures and petrophysical rock fabrics in the core and calibration of the rock fabrics to wireline logs, (3) constructing a sequence stratigraphic framework from core descriptions and wireline log correlations, (4) determining the degree of conformance between depositional textures and diagenetic overprint, (5) constructing rock-fabric layers and mapping the rock-fabric patterns within the layers, and (6) linearly interpolating averaged data or stochastically distributing petrophysical properties within the rock-fabric layers. This method has been used to construct a reservoir model in the Seminole San Andres field, West Texas, and the results are presented next.

8.6.2
Seminole Field and Geological Setting

The Seminole reservoir is located in Gaines County, Texas and lies within the Permian Basin on the northern Central Basin Platform immediately south of the San Simon Channel (Fig. 23). The reservoir produces from the San Andres Formation, Guadalupian age. Seismic data suggest that Seminole field is one of several isolated platforms built during the lower San Andres that became linked with the rest of the San Andres platform during progradation of the upper San Andres.

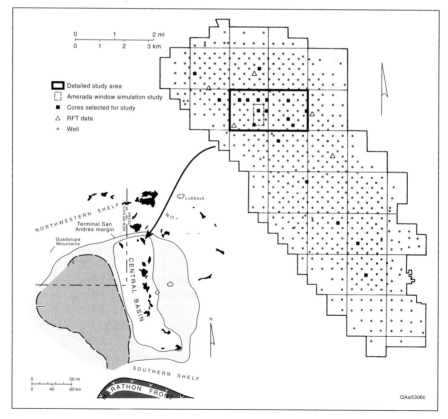

Fig. 23. Location map of the Seminole field in the Permian Basin, West Texas, and the location of the two-section study area (bold outline) (Lucia et al. 1995)

The field covers approximately 23 square miles and contains over 600 wells. Discovered in 1936, the reservoir is a solution-gas drive type with a small initial gas cap. Original oil in place is estimated to be 1100 MSTB (Galloway et al. 1983). Waterflooding was initiated in 1970 using alternating rows of 160-acre inverted nine-spot patterns. Infill drilling occurred in 1976, converting the pattern to a mixed 80- and 160-acre inverted nine spot. A second infill program took place in 1984–1985, converting the pattern to a 80-acre inverted nine spot. CO_2 flooding began in 1985.

8.6.3
Rock-Fabric-Petrophysical Relationships

Core analysis data available for this study included whole-core porosity and permeability measurements, and a limited number of capillary pressure curves and relative permeability curves. SSAU well 2505 was selected for detailed rock fabric analysis, which was a poor choice because it was determined later that the core analysis was in error. The error was uncovered because the rock-fabric-specific porosity-permeability transforms deviated from the generic cross plots presented by Lucia (1995), and because total point-count porosity from thin sections was much higher than core porosity.

To check the accuracy of the core analysis, three core plugs were drilled from each of 12 whole-core samples. Although the whole-core samples had been cleaned of hydrocarbons before they were initially analyzed, the plugs were cleaned again. The permeability of the recleaned samples compared well with the original values, but the porosity of the recleaned samples was 2 porosity units higher than the original values (Fig. 24). This suggests that the core was improperly cleaned before whole core analysis was done, resulting in low porosity measurements. Permeability values were unaffected because the large pores carry most of the flow and were well cleaned initially. Only the small pores still contained residual oil (Fig. 24).

The interparticle porosity, permeability, rock-fabric cross plot using data from the cleaned plugs is in agreement with the generic plot (Fig. 25). The recleaned samples are dolograinstone (class 1), grain-dominated dolopackstone (class 2), and mud-dominated dolostone with 20 to 25 mm dolomite crystals (class 2.5 because it falls on the line between classes 2 and 3). None of the recleaned samples are class 3. This is a very limited data set on which to base rock-fabric-specific porosity-permeability transforms. Therefore, a technique integrating core permeability values with rock-fabric classes and interparticle porosity calculated from wireline logs was used to develop rock-fabric-specific porosity-permeability transforms. This technique will be discussed in the section on wireline log methods.

The Seminole San Andres reservoir produces from anhydritic dolomite and contains five principal rock fabrics; (91) dolograinstone (class 1), (2) fine to medium crystalline grain-dominated dolopackstone (class 2), (93) fine crystalline mud-dominated dolostone (class 3), (4) medium crystalline mud-dominated

Fig. 24. A Cross plot of porosity values of plug samples taken from the whole core samples and recleaned versus whole-core porosity values. Whole-core porosity is too low by 0 to 4 porosity percent. B Graphic display showing the range of permeability in selected whole-core samples. Values from three plugs about 1 in. apart are compared with original whole-core permeability values

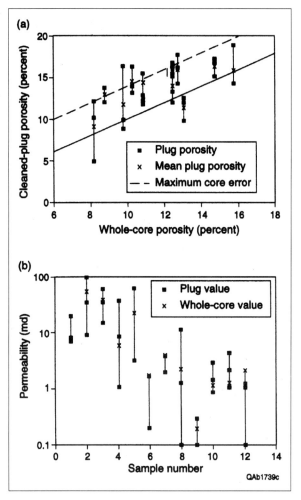

dolostone (class 2), and (5) separate-vug (moldic and intrafossil) dolostones (Fig. 26).

Dolograinstone is the least common fabric in the two section study area. In this study, the term grainstone is used in the strict sense as a grain supported texture with no intergrain lime mud. Grain types are normally either peloids or fusulinids with occasional ooids. The generic porosity-permeability cross plot (Fig. 27) shows that this fabric has the highest flow potential. However, it is often cemented tight with anhydrite. Generic capillary pressure data show that this fabric is characterized by the lowest water saturation (Fig. 27).

Grain-dominated dolopackstones are very common in the Seminole reservoir. They are often misidentified as grainstones because they are grain supported and may have very little intergrain mud. Peloids and fusulinids are the most common grain type. The generic porosity-permeability cross plot shows this

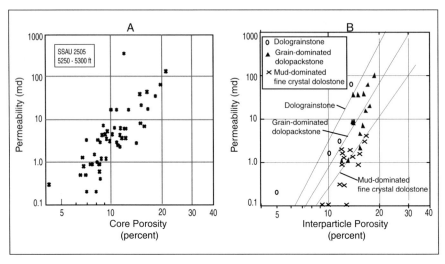

Fig. 25. Porosity, permeability, rock-fabric transforms from core plugs, well 2505. A Cross plot of porosity versus permeability for grain-dominated dolopackstone and medium crystalline mud-dominated dolostones in the lower portion of the reservoir. The distribution conforms to petrophysical class 2. B Cross plot of interparticle porosity versus permeability using data from recleaned core plugs showing higher permeability for grain-dominated dolopackstones than mud-dominated dolostones. The transforms conform to the generic petrophysical class fields of Lucia (1995) (Lucia et al. 1995)

fabric to be in petrophysical class 2 with slightly less flow potential than grainstone (Fig. 27). Generic class 2 capillary pressure curves characteristically have higher water saturations than class 1 grainstones.

Medium crystalline mud-dominated dolostones are very common in the lower portions of the reservoir, and they are characterized by a class 2 porosity-permeability transform (Fig. 25). As can be seen by the generic porosity-permeability cross plot, the larger dolomite crystal size elevates mud-dominated fabrics from class 3 into petrophysical class 2 because the intercrystal pore size is larger than interparticle pore size in micrite (Fig. 27).

Fine crystalline mud-dominated-packstones, wackestones, and mudstones are common in the upper portions of the reservoir. Common grain types are peloids, mollusk fragments, and fusulinids. The generic porosity-permeability cross-plot places this fabric in petrophysical class 3 with the lowest flow potential. Generic capillary pressure relationships suggest that this class is characterized by the highest water saturation (Fig. 27).

Grain molds and intrafossil pores are the principal separate-vug types and consist of (1) dissolved peloids, mollusk fragments, and fusulinids, and (2) intrafusulinid pores. An interval containing separate-vug porosity in SSAU 2309 well was sampled to investigate the effect of separate vugs on permeability. As is expected from generic relationships, the presence of this pore type reduces permeability over what would be expected if all the porosity were interparticle (Fig. 28).

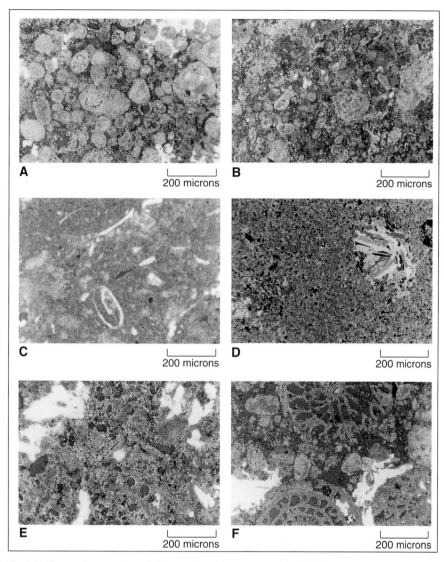

Fig. 26. Photomicrographs of thin sections impregnated with blue dye illustrating rock fabrics (white areas are anhydrite). **A** Dolograinstone with intergrain pore space; **B** grain-dominated dolopackstone with intergrain pore space and intergrain dolomitized micrite; **C** fine crystalline dolowackestone; **D** medium crystalline dolowackestone (cross polarizers); **E** separate-vug (moldic) porosity in grain-dominated dolopackstone; and **F** separate-vug (intrafusulinid) porosity in grain-dominated dolopackstone

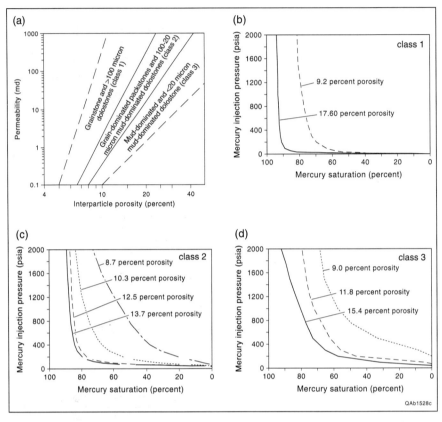

Fig. 27. Generic porosity, permeability, rock-fabric cross plot and rock-fabric specific capillary pressure plots (Lucia, 1995)

Fig. 28. Cross plot of total porosity and permeability showing moldic grain-dominated packstones with lower permeability than expected for class 2 grain-dominated dolopackstones with high porosity (SSAU 2309 well)

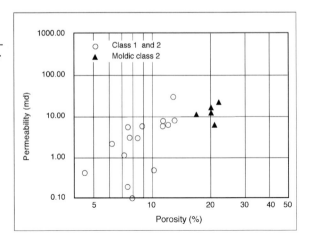

8.6.4
Vertical Succession of Rock Fabrics and Depositional Textures from Core

Descriptions of the vertical succession of rock fabrics for petrophysical quantification and depositional textures for sequence stratigraphic analysis are closely linked and are best done in concert. It is common to use core slab descriptions for sequence stratigraphic analysis, but thin sections are commonly needed to make the fine distinctions between (1) grainstone, grain-dominated packstone, and mud-dominated fabrics, (2) inter- and intraparticle pore space, and (3) fine, medium, and large dolomite crystal sizes. This difference in scale of observation commonly results in some differences in fabric descriptions which must be resolved as part of the process of integrating geological descriptions and petrophysical data (Fig. 29).

The depositional textures and rock fabrics are stacked vertically into two basic depositional cycles: subtidal and tidal-flat capped. The depositional textures are overprinted by cementation, compaction, some selective dissolution, reflux dolomitization, and evaporate mineralization to form the five basic rock-fabrics listed above. Tidal-flat-capped cycles are concentrated in the dense upper 300 ft of the San Andres Formation. The lithology is a fine crystalline anhydritic dolostone that forms the upper seal for the reservoir and typically has less than 5% porosity and 0.1 md permeability. Within the reservoir a few of the upper cycles have discontinuous thin tidal-flat caps.

The most common depositional cycle in the reservoir is the subtidal, upward-coarsening dolomitized cycle composed of a lower mud-dominated unit overlain by a grain-dominated unit. Most subtidal cycles are capped by a mud- or grain-dominated packstone, but a few cycles have grainstone caps.

The reservoir has been divided into 12 high-frequency cycles. The lower three HFC (cycles 10, 11, and 12) are part of the outer-ramp facies tract, and are fusulinid rich 40- to 50-ft-thick units containing low porosity fusulinid-peloid mud-dominated dolostones coarsening upwards into more porous crinoid-fusulinid-peloid grain-dominated dolopackstones (Fig. 29). Cycle 12 is only partially described. Cycles 11 and 10 are thick cycles suggesting high accommodation typical of the TST. The upper nine HFC (cycles 1, 2, 3, 4, 5, 6, 7, 8, and 9) contain fewer fusulinids and record progradation of ramp-crest and inner ramp facies. HFCs are typical upward-shallowing cycles with basal mudstones and wackestones grading upward into grain-dominated packstones and grainstones (Fig. 29). Cycles 5 and 6 are the first tidal-flat-capped cycles encountered in the vertical succession. A fusulinid wackestone overlies cycle 5 suggesting a significant flood back and deepening event. Cycle 1 is capped by a thick interval of grain-dominated sediment which is in turn overlain by the upper 300 ft of peritidal deposits. This major facies offset probably marks a sequence boundary.

Rock-fabric descriptions differ from descriptions of depositional textures in that they include effects of diagenesis; rock fabrics describe the current conditions. The diagenetic overprints on the depositional textures in the Seminole Field are compaction, cementation, selective dissolution, reflux dolomitization,

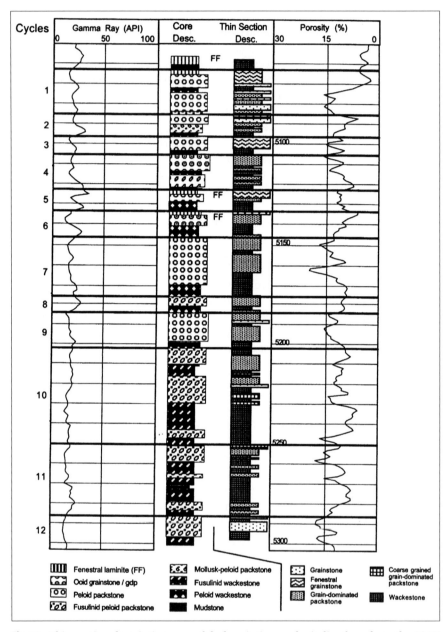

Fig. 29. Thin section description, core-slab description, and wireline logs from the SSAU 2505 well showing, (1) vertical succession of rock-fabric facies, (2) high-frequency cycles, and (3) detailed thin section descriptions . (Lucia et al., 1995)

and anhydrite emplacement. Compaction and cementation were the dominant diagenetic processes early in diagenetic history and the products conform to depositional textures. Mud-dominated sediments compact more than grain-dominated sediments, resulting in cycles characterized by an upward increase in porosity and permeability (Fig. 30a). Pore space in grainstones was partially occluded by early marine cement.

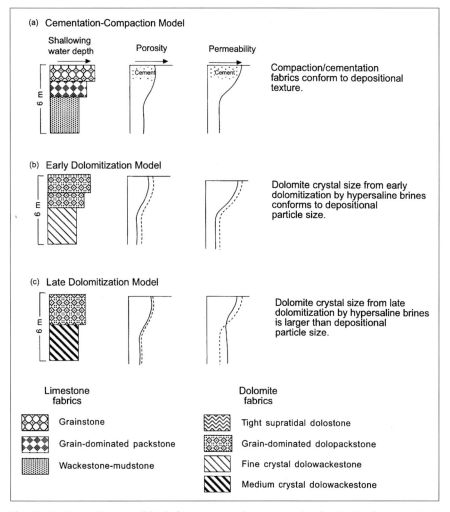

Fig. 30. Geologic history of high-frequency cycles present in the Seminole reservoir. A Shoaling-upward cycle showing porosity and permeability profiles resulting from compaction-cementation diagenesis. B Shoaling-upward cycle showing early dolomitization, reduction of porosity and permeability due to overdolomitization and anhydrite precipitation. C Shallowing-upward cycle showing later dolomitization and improved permeability in dolomitized mud-dominated dolostone due to increase in particle size

Reflux dolomitization and anhydrite emplacement followed limestone compaction and cementation diagenesis. There is little to suggest conformance between anhydrite distribution and depositional texture, and dolomitization probably resulted in a modest loss of porosity in all depositional facies due to overdolomitization (Lucia and Major 1994). Conformance of dolomite textures, however, depends largely upon the dolomite crystal size.

Dolomite crystal size increases with depth from 10–20 mm to 50–100 mm (Fig. 31). Dolomite crystals in grain-dominated fabrics are larger than in mud-dominated fabrics. Mud-dominated fabrics in the upper portion of the reservoir are characterized by fine crystalline dolostones, and the dolomite fabrics mimic the precursor limestone fabrics. These fabrics are petrophysical class 3. In the lower portion of the reservoir, the dolomite crystal size of mud-dominated dolostone is typically between 20 and 100 mm, and the petrophysical characteristics of the mud-dominated dolostone no longer mimic the precursor limestone. The increase in particle size from fine to medium results in an increase in pore size and improvement in the petrophysical properties. Assuming early hypersaline reflux dolomitization from peritidal sediments in the upper 300 ft, the fine crystalline characteristics of the upper portion result from early dolomitization relative to the time of deposition (Fig. 30b), whereas the medium crystalline characteristics of the lower subtidal cycles result from later dolomitization relative to the time of deposition (Fig. 30c).

Fig. 31. Cross plot of depth versus dolomite crystal size for mud-dominated and grain-dominated dolostones in well 2505 showing a somewhat cyclic but systematic increase in crystal size with depth

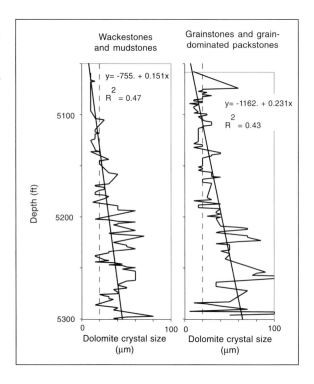

8.6.5
Sequence Stratigraphy

The San Andres Formation is some 1500 ft thick. An outcrop study of the San Andres Formation, Algerita Escarpment, Guadalupe Mountains, New Mexico and Texas has shown that the lower 700 ft of the San Andres contains skeletal grainstone and packstone and an open-marine fauna. This interval is interpreted to be a transgressive systems tract of the lower San Andres composite sequence of latest Leonardian age (Kerans et al. 1993, 1994). The highstand systems tract of the upper San Andres composite sequence is represented by (1) a lower 300 ft of fusulinid wackestones and packstones, (2) 150 ft of upward-shallowing peloidal shallow subtidal to peritidal cycles, and (3) an upper 350 ft of largely anhydritic peritidal deposits.

The vertical facies succession from 11 cores in the upper composite sequence were described in detail and form the basis for constructing a chronostratigraphic sequence stratigraphic framework for the Seminole field. Through an interactive process of core description and correlation, 12 high-frequency cycles (HFC) were confidently identified in cores and correlated with wireline logs (Kerans et al. 1993). The vertical facies succession is similar to that described by the San Andres outcrop suggesting that the Seminole reservoir also lies in the highstand systems tract of the upper San Andres composite sequence. Two high-frequency sequences are suggested by the vertical facies succession and lateral continuity of facies. Cycles 11 and 10 appear to be retrogradation facies and cycles 9-5 appear to be progradational facies with a sequence boundary at the top of cycle 5. Cycle 4 marks a deepening event because it contains abundant fusulinids, a well-established key indicator of water depth. The facies offset above cycle 1 suggests a second sequence boundary.

8.6.6
Wireline Log Analysis

8.6.6.1
Introduction

Only 11 of the 58 wells in the two-section study area have core information. Therefore, methods of extracting accurate values of porosity, saturation, and permeability from wireline log data are needed to increase the density of data before interpolation between wells is attempted. Of the 58 wells, 33 have compensated neutron, density, acoustic, dual laterolog, and microfocused logs. Of these logs, the acoustic and resistivity logs respond to rock fabric elements of particle size and sorting, and separate-vug porosity. Therefore, these textural elements can be extracted from wireline data using various multivariant analysis methods.

8.6.6.2
Porosity

Total porosity is calculated by using CNL, density, and acoustic logs. Neutron-density cross-plot porosity is in good agreement with core porosity in well 2505. However, the core porosity has been shown to be too low by several porosity units (PU). Also, a cross plot of acoustic Δt and neutron-density porosity has an intercept of 48.5 ms/ft, the velocity of limestone, resulting in erroneous grain density values.

The correct lithology was obtained by using neutron, density, and acoustic logs to calculate porosity. A pseudo-fluid transit time of 150 µs/ft was used because the transit time of water (189 µs/ft) resulted in unreasonably low porosity values. Using three porosity tools and a pseudo fluid transit time of 150 µs/ft, porosity values were obtained that agree with the recleaned porosity samples and are slightly higher than the original core porosity values. The cross plot of the 3-log porosity values and Δt results in an intercept of 46 µs/ft, an acceptable value for anhydritic dolomite (anhydrite is 50 µs/ft, dolomite is 44 µs/ft).

8.6.6.3
Separate-Vug Porosity

The pseudo-fluid transit time value can be related to separate vug porosity (Lucia and Conti 1987). Lucia and Conti assume that the Wyllie Time Average equation is correct for carbonate rocks with equal volumes of separate vugs. They develop an empirical equation relating separate-vug porosity to the departure from the ideal, an approach that has been used by a number of authors (Wang and Lucia 1993). In a similar manner, an empirical equation relating separate-vug porosity to total porosity and Δt has been established for the Seminole field (see equation below; Fig. 32). The equation is similar to Lucia and Conti's, but it is shifted due to the difference in velocity of limestone and anhydritic dolomite.

$$\phi_{sv} = (2.766 \times 10^{-4}) \times (10^{[-0.1526(\Delta t - 141.5\phi)]})$$

With this equation, total porosity can be divided into interparticle porosity and separate-vug porosity by subtracting total porosity from separate-vug porosity.

8.6.6.4
Water Saturation

Water saturations were calculated using the Archie Equation and a variable m. A value of 2 was used for the saturation exponent (n) and the resistivity of the original connate water is given as 0.2 ohm m. The cementation factor (m) was calculated from the ratio of separate-vug porosity (calculated from the acoustic-porosity relationship) to total porosity (3-log porosity) using a transform shown below developed using data from Lucia (1983) and Lucia and Conti (1987).

$$m = 2.14(\phi sv/\phi t) + 1.76$$

Fig. 32. Relationship between transit time, total porosity, and separate-vug porosity in Amerada Hess No. 2505 well (Lucia et al. 1995)

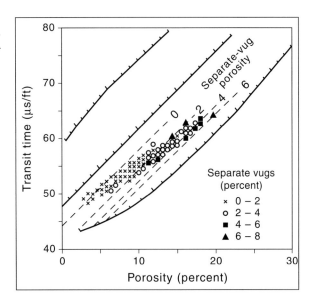

8.6.6.5
Particle Size and Sorting

The combination of interparticle porosity, particle size, and sorting characterizes interparticle pore-size distribution. Water saturation is also a function of pore-size distribution and can be used to estimate particle size and sorting in wells that have no cores. In the Seminole reservoir, particle size and sorting was calculated using a relationship between porosity, saturation, particle size and sorting, and reservoir height. Using data from cycles 1-9 in SSAU 2505, water saturations and porosity from log calculations were compared with rock-fabrics described from thin sections (Fig. 33). A cross plot of average porosity and water

Fig. 33. Relationship between water saturation, porosity, and rock-fabric/petrophysical class. The porosity and saturation values have been averaged within rock-fabric units in cycles 1-9, well 2505 (Lucia et al. 1995)

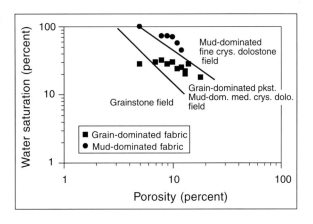

saturation shows that fine crystalline mud-dominated dolostones (petrophysical class 3) can be separated from grain-dominated dolopackstones and medium crystalline mud-dominated dolostones (petrophysical class 2). Too few grainstone fabrics are found in well 2505 for calibration, but, if present, should plot below the class 2 field.

The limitations of using water saturation to determine particle size and sorting are listed below.

1. The method does not work in 100% water saturated intervals. The Seminole logs used to construct the model are all within the oil column.
2. The height above the zero-capillary-pressure elevation (herein called reservoir height) must be accounted for. Only intervals well above the zero-capillary-pressure-elevation were used in the Seminole study.
3. The method will not work in zones that have been water flooded. In the Seminole field, wells that were completed water free are considered to have original water saturation values and original water resistivity. The method was used only in these wells. Wells that were completed producing water and oil are considered to contain water-flooded zones.

Figure 34 shows the point data for three wells and the fabric fields that define classes 1, 2 and 3. The classes were defined by core descriptions. Equations that describe the boundaries between the fabric fields are listed below:

Class 1/2 boundary = Sw = $(6.522 \times 10^{-3}) \times (\phi^{-1.401})$,

Class 2/3 boundary = Sw = $(3.05 \times 10^{-2}) \times (\phi^{-0.981})$.

8.6.6.6
Permeability

Calibrating core permeability with wireline logs is a difficult task. Historically, permeability has been estimated using a single porosity-permeability transform. In this study we have used rock-fabric-specific porosity-permeability transforms to estimate permeability. Because the porosity values in the core from well 2505 are suspect, the porosity permeability transforms were determined by cross plotting log calculated interparticle porosity and core permeability. Critical steps in this procedure are averaging the core data to match the volume of rock sampled by porosity logs, and depth shifting core data to match wireline log data. Even when done properly the results are more uncertain than cross-plotting core porosity, permeability and rock-fabric data measured on the same sample. Figure 34 shows cross plots of log-calculated interparticle porosity and core permeability for the three classes from three cored wells. Although there is considerable scatter in the data, the resulting transforms show an improvement in permeability from class 3 to class 1, a trend which is consistent with generic transforms. It should be noted that these transforms, especially class 3, do not coincide exactly with generic fields based solely on core data. This results from using log-calculated interparticle porosity values rather

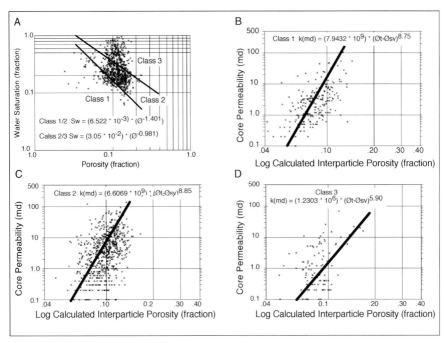

Fig. 34. Permeability calculation method. **A** Relationship between water saturation, porosity, and rock-fabric/petrophysical class using point data from well log calculations above transition zone in wells 2505, 2309, and 2504 (lines show boundaries between classes 1, 2, and 3 used in this study). **B** Cross plot of core permeability and log calculated interparticle porosity for class 1 points. **C** Cross plot of core permeability and log calculated interparticle porosity for class 2 points. **D** Cross plot of core permeability and log calculated interparticle porosity for class 3 points. Transforms used in this study to calculate permeability are shown. Note that the transforms for classes 1 and 2 are not very different

than core porosity values, and the problems of sampling different volumes of rock.

The permeability transforms used to construct the reservoir model are presented below:

$$\text{Class 1 } k(md) = (7.9432 * 10^9) \times (\phi_t - \phi_{sv})8.75,$$

$$\text{Class 2 } k(md) = (6.6069 * 10^9) \times (\phi_t - \phi_{sv})8.85,$$

$$\text{Class 3 } k(md) = (1.2303 * 10^6) \times (\phi_t - \phi_{sv})5.90.$$

8.6.7
Reservoir Model Construction

There are four steps in the construction of a rock-fabric layered model: (1) construction of the HFC framework, (2) identification of petrophysically significant rock-fabric facies within each well, (3) construction of rock-fabric layers, and (4) averaging petrophysical properties within rock-fabric facies and linear interpolation of properties between wells. The HFC framework has been constructed and petrophysically significant rock-fabric facies identified in cored and uncored wells. The rock-fabric flow layers are constructed by correlating the rock fabric flow units between wells within the HFC framework.

Outcrop studies describe thin dense discontinuous mud layers that are important barriers to vertical flow. Core descriptions suggest that these dense mud layers are present in the Seminole field as well. However, wireline logs average data over several feet so that the low porosity of these dense layers is averaged with the porosity of surrounding beds. Therefore, the discontinuous thin dense layers are inserted into the model based on core data and an outcrop model.

8.6.7.1
Rock-Fabric Flow Units

The term flow units has been used in many different ways. It is defined here as a rock-fabric facies in which the petrophysical properties are near randomly distributed. Rock-fabric facies are defined by limestone textures whereas petrophysical properties are defined by integrating rock-fabric classes into wireline log analysis. Petrophysical classes include the effects of diagenesis, whereas rock-fabric facies are defined by limestone textures. When diagenetic products do not mimic limestone textures, petrophysical classes and limestone textures will not conform.

In the Seminole field, this lack of conformance is manifest in cycles 10 through 12 where mud-dominated fabrics typically have medium size dolomite crystals and are class 2 rather than class 3. Log analysis shows these cycles to consist of a single class 2 rock fabric, whereas core descriptions show cycles 10 through 12 to consist of a lower medium crystalline dolowackestone and an upper medium crystalline mud-dominated dolopackstone. The packstone has higher porosity than the wackestone, a difference that is probably inherited from differential compaction of the precursor limestone. The porosity difference is also reflected in the permeability because the same class 2 rock-fabric-specific permeability transform is used (Fig. 23). Therefore, the rock-fabric flow units in cycles 10, 11, and 12 are based on porosity differences because they can be related to depositional texture (Fig. 35).

As shown in Fig. 35, cycle 11 contains 4 rock fabric layers defined by (1) an upper grain-dominated packstone, (2) a lower wackestone with two thin beds of grain-dominated packstone, and (3) a dense mudstone. Using the class 2 porosity-permeability transform for all four flow layers results in an excellent match between core and calculated permeability values.

Above cycle 10, the dolomite crystal size tends to mimic the precursor limestone texture, and petrophysical rock-fabric classes 1, 2, and 3 conform with limestone textures grainstone, grain-dominated packstone, and mud-dominated fabrics. As shown in Fig. 34, cycles 4, 3, and 2 are shallowing upward cycles

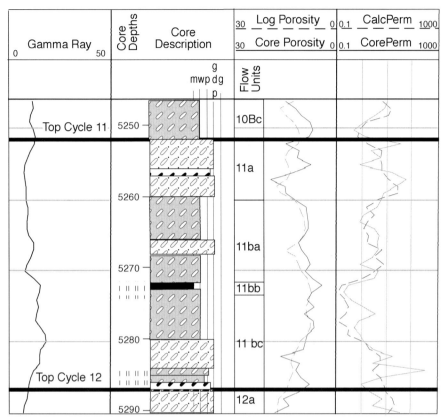

Fig. 35. Cycle 11 core description, rock-fabric flow units, core and log-calculated porosity and permeability, and gamma ray log, Amerada Hess SSAU 2505 well. Rock-fabric flow unit 11a is distinguished by higher porosity values, and rock-fabric flow unit 11bb is based in a tight mudstone layer from the core description. Permeability calculations are based on a single class 2 porosity-permeability transform (Lucia et al. 1995)

and each contains two rock-fabric flow units: a lower class 3, mud-dominated unit, and an upper class 2 or 1, grain-dominated unit. Three rock-fabric-specific porosity-permeability transforms were used, and the results are in excellent agreement with core permeability (Fig. 36).

8.6.7.2
Reservoir Model

Rock-fabric flow units are defined from wireline log calculations in each well. In cycles 12-10 they are based on porosity differences within each cycle, and in cycles 9-1 they are based on petrophysical classes. Dense thin mud layers are superimposed on these flow units to complete the definition of flow units. The flow

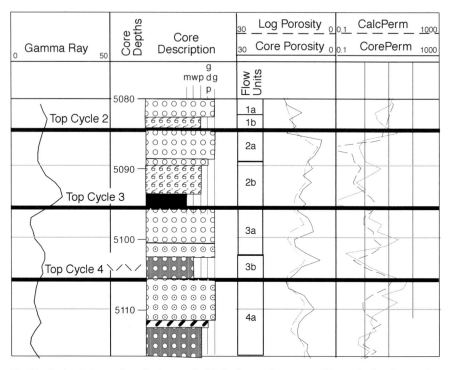

Fig. 36. Cycles 2-4 core description, rock-fabric flow units, core and log-calculated porosity and permeability, and gamma-ray log, Amerada Hess SSAU 2505. Rock-fabric flow units are based on petrophysical classes from wireline log calculations and conform well with rock-fabric facies from core descriptions. Permeability is calculated using three rock-fabric porosity-permeability transforms (Lucia et al. 1995)

units are correlated between wells to form the rock-fabric flow layers. Because most reservoir simulation programs do not allow for discontinuous layers, all flow layer boundaries must be continuous within the model. This results in one rock-fabric facies containing more than one rock-fabric flow layer. Where this occurs, the same facies-average petrophysical value is used in each layer within the rock-fabric facies.

The resulting reservoir model consists of 41 rock-fabric layers (Fig. 37). Rock-fabric facies change laterally within the rock-fabric layers. There is little lateral rock-fabric variability in the transgressive system tracts (cycles 12-10), whereas more lateral variability is found in the prograding systems tracts (cycles 9-5, 3-1). No sharp lateral boundaries are placed between the facies because no sharp boundaries have been found in analog outcrops on the scale of 1000 ft.

For each well, the petrophysical data within each rock-fabric facies are averaged, assigned to a rock-fabric layer, and interpolated between wells. The layers and average petrophysical data are input into a fluid flow simulator, and the resulting sim-

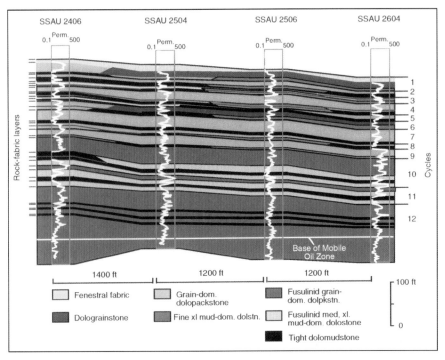

Fig. 37. Cross section illustrating the distribution of rock-fabric facies and rock-fabric flow layers within the high-frequency cycle framework for a portion of the Seminole San Andres Unit

ulation model is shown in Fig. 38. The simulation model contains 41 rock-fabric flow layers within the 12 high-frequency cycles. Petrophysical properties vary laterally on the scale of 1000 ft and vertically on the scale of feet or tens of feet. This model was used in flow simulations to determine the location of remaining oil.

8.6.8
Simulation Results

The rock fabric method described herein results in a 3-D image of petrophysical properties that is comparable with outcrop descriptions. It preserves not only the cycle-based geological framework but also the architecture of petrophysically based rock-fabric facies. The problem of scale averaging is minimized by averaging petrophysical data within rock-fabric facies and using rock-fabric layers for constructing the simulation model. The use of rock-fabric specific permeability transforms preserves high and low values better than the commonly used single transform, and calculation of separate-vug porosity eliminates overestimation of permeability in zones of high separate-vug porosity. In addition, rock-fabric simulation models allow for easy inclusion of rock-fabric specific relative permeability curves in the model (Wang et al. 1994).

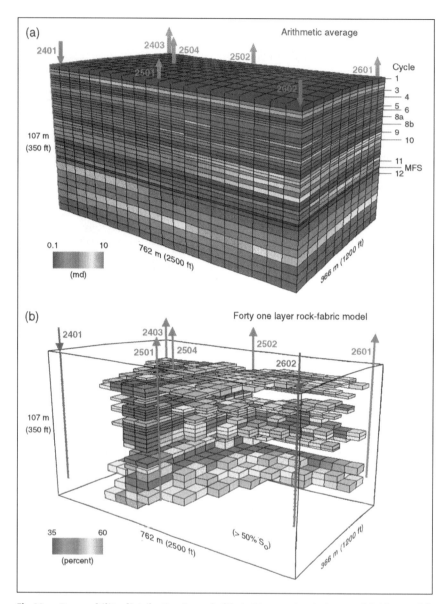

Fig. 38. a Permeability distribution in rock-fabric 80-acre simulation model. The model contains 41 rock-fabric layers. **b** Simulation results showing distribution of greater than 50% oil saturation after 17 years of waterflooding

The 41-layer simulation model was used to simulate production from an 80 acre area within the two section study area. The resulting image of the remaining oil saturation (Fig. 38) shows layers of bypassed oil. Recovery efficiency from this type of reservoir is typically 35% of the original oil in place from primary production and conventional waterflooding, and the simulation results shows the remaining 65% to be located in bypassed low-permeability layers. This is a more realistic image and will provide more accurate performance predictions of proposed recovery programs (Wang et al. 1994; Lucia et al. 1995).

References

Barber AJ, George CJ, Stiles LH, Thompson BB (1983) Infill drilling to increase reserves – actual experience in nine fields in Texas, Oklahoma, and Illinois. J Pet Technol August 1983: 1530–1538

Fogg GE, Lucia FJ (1990) Reservoir modeling of restricted platform carbonates: geologic/geostatistical characterization of interwell-scale reservoir heterogeneity, Dune Field, Crane County, Texas. The University of Texas at Austin, Bureau of Economic Geology, Report of Investigation No 190, 66 pp

Galloway WE, Ewing TE, Garrett CM, Tyler N, Bebout DG (1983) Atlas of major Texas oil reservoirs. The University of Texas at Austin, Bureau of Economic Geology, 139 pp

George CJ, Stiles LH (1978) Improved techniques for evaluating carbonate waterfloods in West Texas. J Pet Technol Nov 1978: 1547–1554

Grant CW, Goggin DJ, Harris PM (1994) Outcrop analog for cyclic-shelf reservoirs, San Andres Formation of Permian Basin: stratigraphic framework, permeability distribution, geostatistics, and fluid-flow modeling. AAPG Bull 78 1: 23–54

Hearn CJ, Ebanks WF Jr, Tye RS, Ranganathan V (1984) Geological factors influencing reservoir performance on the Hartzog Draw field, Wyoming. J Pet Technol Aug 1984: 1335–1344

Hovorka SD, Nance HS, Kerans C (1993) Parasequence geometry as a control on porosity evolution: examples from the San Andres and Grayburg formation in the Guadalupe Mountains, New Mexico. In: Loucks RG, Sarg JF (eds) Carbonate sequence stratigraphy: recent developments and applications. AAPG Mem 57: 493–514

Kerans C, Lucia FJ, Senger RK, Fogg GE, Nance HS, Hovorka SD (1993) Characterization of facies and permeability patterns in carbonate reservoirs based on outcrop analogs. The University of Texas at Austin, Bureau of Economic Geology, final report prepared for the Assistant Secretary for Fossil Energy, US Department of Energy, under contract no DE-AC22-89BC14470, 160 pp

Kerans C, Lucia FJ, Senger RK (1994) Integrated characterization of carbonate ramp reservoirs using Permian San Andres Formation outcrop analogs. AAPG Bull 78, 2: 181–216

Lucia FJ (1983) Petrophysical parameters estimated from visual descriptions of carbonate rocks: A field classification of carbonate pore space. J Pet Technol 35, 3: 629–637

Lucia FJ (1995) Rock-fabric/petrophysical classification of carbonate pore space for reservoir characterization. AAPG Bull 79, 9: 1275–1300

Lucia FJ, Major RP (1994) Porosity evolution through hypersaline reflux dolomitization. In: Purser BH, Tucker ME, Zenger DH (eds) Dolomites, a volume in honor of Dolomieu. Int Assoc Sedimentol Spec Publ 21: 325–341

Lucia FJ, Kerans C, Senger RK (1992) Defining flow units in dolomitized carbonate-ramp reservoirs. Soc Petroleum Engineers Techn Conf, Washington D.C., SPE 24702, pp 399–406

Lucia FJ, Kerans C, Wang FP (1995) Fluid-flow characterization of dolomitized carbonate-ramp reservoirs: San Andres Formation (Permian) of Seminole field and Algerita Escarpment, Permian Basin, Texas and New Mexico. In: Stoudt EL, Harris PM (eds) Hydrocarbon reservoir characterization: geologic framework and flow unit modeling. SEPM (Society for Sedimentary Geology), SEPM Short Course 34: 129–153

Lucia RJ, Conti RD (1987) Rock fabric, permeability, and log relationships in an upward-shoaling, vuggy carbonate sequence. The University of Texas at Austin, Bureau of Economic Geology, Geological Circular 87–5, 22 pp

Senger RK, Lucia FJ, Kerans C, Ferris MA, Fogg GE (1993) Dominant control on reservoir-flow behavior in carbonate reservoirs as determined from outcrop studies. In: Linville B, Burch-

field TE, Wesson TC (eds) Reservoir characterization III. Proc 3rd Int Reservoir Character-
    ization Techn Conf, Tulsa, Nov 1991. PennWell Books, Tulsa, Oklahoma, pp 107–150
Wang FP, Lucia FJ (1993) Comparison of empirical models for calculating the vuggy porosity
    and cementation exponent of carbonates from log responses. The University of Texas at
    Austin, Bureau of Economic Geology, Geological Circular 93–4, 27 pp
Wang FP, Lucia FJ, Kerans C (1994) Critical scales, upscaling, and modeling of shallow-water
    carbonate reservoirs. Soc Petroleum Engineers Tech Conf, Midland, Texas, SPE 27715, pp
    765–773
Wang FP, Lucia FJ, Kerans C (1998) Integrated reservoir characterization study of a carbonate
    ramp reservoir: Seminole San Andres Unit, Gaines County, Texas. SPE Reservoir Evalua-
    tion & Engineering 1, 3: 105–114

Subject Index

Printing: Saladruck, Berlin
Binding: Buchbinderei Lüderitz & Bauer, Berlin